Writing, Geometry and Space in Seventeenth-Century England and America

For many cultural historians the early modern map has come to mark the threshold of modernity, cutting through the layered customs of Medieval parochialism with its clean, expansive geometries. Re-thinking the role played by mathematics and cartography in the English seventeenth century, this book argues that the cultural currency of mathematics was as unstable in the period as that of England's controversial enclosures and plantations. Reviewing evidence from a wide range of literary and scientific texts, Jess Edwards suggests that its unstable currency rendered mathematics necessarily rhetorical: subject to constant re-negotiation. Yet he also finds a powerful flexibility in this weakness. Mathematized texts from masques to maps negotiated a contemporary ambivalence between Calvinist asceticism and humanist engagement. Their authors promoted themselves as artful guides between virtue and profit; the study and the marketplace.

Jess Edwards is Principal Lecturer in English at Manchester Metropolitan University.

Routledge Studies in Renaissance Literature and Culture

Writing, Geometry and Space in Seventeenth-Century England and America

Circles in the sand

Jess Edwards

Routledge
Taylor & Francis Group

LONDON AND NEW YORK

First published 2006
by Routledge
2 Park Square, Milton Park, Abingdon, Oxon OX14 4RN

Simultaneously published in the USA and Canada
by Routledge
711 Third Avenure, New York, NY 10017

Routledge is an imprint of the Taylor & Francis Group

First issued in paperback 2016

Typeset in Garamond by Wearset Ltd, Boldon, Tyne and Wear

British Library Cataloguing in Publication Data
A catalogue record for this book is available from the British Library

Library of Congress Cataloging in Publication Data
A catalog record for this book has been requested

ISBN13: 9-78-0-415-32341-3 (hbk)
ISBN13: 9-78-1-138-81005-1 (pbk)

To Mary and Leo: for love, support and patience

Contents

Figures

Acknowledgements

This book began an embarrassingly long time ago as an MA dissertation at University College London, evolving into a doctoral thesis at the University of Sussex. It's a pleasure to be able to thank those who supported and inspired me at both institutions. I received wonderful teaching as an undergraduate at UCL. Without Danny Karlin's encouragement, however, I wouldn't have gone any further. Ivy Schweitzer, who spent a term on loan to UCL from Dartmouth during my MA year, was an extraordinary inspiration and remains my ideal as a teacher. She's the reason I'm an early modernist. At Sussex my supervisor Stephen Fender steered me patiently through endless digressions and delusions and Peter Nicholls was also a great friend and support. Alongside my research I taught first as a Visiting Lecturer at Roehampton Institute, then as a Lecturer at the University of North London, now London Metropolitan University. I owe thanks to Janet Beer and Kim Reynolds for their faith and mentorship during my Roehampton years. To Carolyn Burdett and Wendy Wheeler at UNL/London Metropolitan I owe particularly vehement gratitude. Without them I wouldn't have got anywhere with this book. As my highly convivial office mate for four years, Guy Westwell kept me sane. And I would also, in the vaguest but most heartfelt of gestures, like to thank the students I taught at UNL/London Metropolitan. There were many moments, particularly in my first year poetry class, when I couldn't imagine teaching being any more rewarding.

Beyond my institutional career I have received help from many quarters. Jerry Brotton, John Gillies, Andrew Gordon, Richard Helgerson, Bernhard Klein and Garrett Sullivan have all influenced this book immensely in their work and their advice. Jurg Glauser and Christian Kiening organized the superb Zurich conference *Text–Bild–Karte* in 2002 which helped me crystallize my thoughts. Robert Appelbaum was an outstanding editor of one of my articles, whose strictures echo as I write. Lisa Hopkins, Peter Jackson, Andrew McRae, Sarah Mills, Miles Ogborn and Matthew Steggle all responded with remarkable generosity to my unsolicited calls for feedback and advice. Two people deserve final and especial mention. Susan Wiseman has encouraged and advised me since my academic career began and gave unstinting support to this project. As my partner, Mary Peace has kept me going through countless crises; as a reader she has applied her characteristic clarity and nuance. None of those I have acknowledged has been burdened with this book in its entire and finished form, and I take the usual responsibility for whatever faults it retains.

In the course of its long gestation, portions of this book have already been published, though in most cases in a form largely unrecognizable here. I am grateful to Ashgate Press, the University of Pennsylvania Press, and the editors of *Studies in Travel Writing, Early Modern Literary Studies* and *New Formations* for permission to adapt and re-use material either forthcoming or already in print. Permissions to reproduce illustrations have been granted by the British Library (Figures 2.1, 3.1, 3.2, 4.1, 5.2, 6.1, 6.2, 7.1, 7.2) and Massachusetts Archives (4.2, 5.1).

1 Introduction

Writing, geometry and space

Dirty or clean?

It is all too easy to think that we understand the power and the charm of the early modern map. In those often magnificent examples of global, regional and even local cartography which have survived the centuries we imagine we perceive the spirit of European artists, rulers, merchants and landlords for the first time in possession of their material environment; bursting the bubble of Medieval parochialism and stretching out to govern a space as limitless as the geometry which framed it. And if we do not celebrate this breaking free from place to space then we mourn it, as the dawn of a new age of panoptic discipline and surveillance.

For many cultural historians the cartographic mathematization of experience is a crucial marker of modernity, and of the revolution in ideas and values that fostered early modern capitalism and imperialism. If arts such as geography embraced mathematical technologies and aesthetics in the late sixteenth and seventeenth centuries they are held to have done so in an empiricist, pragmatic and mercantile spirit, treating the material world as so much dead matter to be cleanly abstracted, partitioned and exploited: subjected to a distinctly modern form of discipline. Graham Huggan sees in the 'reinscription, enclosure and hierarchization of space' executed by the post-Renaissance map, 'an analogue for the acquisition, management and reinforcement of colonial power' (Huggan 1989: 115). Samuel Edgerton finds in the orthogonal grid common to new forms of cartographic projection and artificial perspective in painting, a 'symbol of cultural expansion' (Edgerton 1983: 22). David Harvey observes that the geometric aesthetic of Ptolemaic cartography made the world in general seem 'conquerable and containable for purposes of human occupancy and action' (David Harvey 1989: 246).

The problem with these judgements is encapsulated in Edgerton's 'symbol' as it is in Huggan's 'analogue'. How do we know what the geometry of early modern maps symbolized or seemed to their early modern makers and users? How can we judge the cultural currency of the early modern map?

Since the 1980s we have become used to regarding early modern cartographies as maps of cultural meaning, rather than simply of material space. But the critical history of early modern cartography that has developed over the past three decades is highly heterogeneous, and is by no means agreed on the ways in which cartography is cultural. The most materialist histories treat maps as 'dirty' entities, locating them in tightly specified local processes and transactions (Brotton 1997). Other scholars have constructed relationships between maps and verbal texts, widely separated in historical time and material space, over the 'cleaner' common ground of symbolism and formal analogy. So is the map clean or dirty? Is it a text that we can read effectively in modern galleries, libraries and classrooms? Is it, on the other hand, a material commodity whose historical significance lies beyond it, in those local transactions and performances in which it was originally engaged?

This question has profound ramifications for the place we give to cartography in early modern culture, and for the way we read or decline to read the map. In this introductory chapter I will attempt an answer, reviewing past strategies and exploring the possibilities for reading cartography with writing. Succeeding chapters will demonstrate that early modern cartography was rhetorical: engendered by and engaged in processes of verbal argument and persuasion. Early modern maps were used not just to represent space but also to negotiate the identity, the legitimacy and the agency of individuals, groups and ventures. As an element in these negotiations they were characteristically entangled in a web of words that all too frequently evaporates in idealist readings and materialist histories of cartography. Neither clean nor dirty, they were often intended as dusty metaphors for the liminal relationship between virtue and profit; knowledge and the world.

From transparent window to thickened text

Older histories of cartography tend to be teleological and idealist.[1] Traditional cartographic history assumes consistent development towards a modern scientific practice founded on the discipline of geometric measurement and projection. It treats individual maps as neutral contributions to a Platonic archive of geographic knowledge. It celebrates the sixteenth-century 'renaissance' of mathematical, and thereby recognizably modern methods of cartographic representation, and the comprehensive seventeenth- and eighteenth-century 'reformation' of cartographic art into cartographic science, charting these developments through a 'canon of great maps' (Edney 1993: 56). It treats the geometric space delineated in early modern maps, if not as a Newtonian absolute category of the world, then as a Kantian absolute category of the human mind.

The story of modern geography as told in this traditional history of cartography is the story of rhetoric displaced by science; of individual skill displaced by duplicable technology; of hierarchies of place displaced by abstract space; and in general of the particular displaced by the universal. E.G.R. Taylor prefaces *Late Tudor and Early Stuart Geography* with the suggestion that the 67 years which her indispensable survey reviews:

> bring the reader to the threshold of that period in which for the first time he feels fully at home: the period in which truth is sought by experiment and observation instead of as formerly by reference to authority and revelation.
>
> (E.G.R. Taylor 1934: v)

Taylor's 'threshold' metaphor suggests that whilst she is documenting a 67-year journey, arrival at the 'home' of modern cartography is as concrete and absolute as crossing a national frontier. Mapping does not really become any more modern once this threshold has been crossed, any more than France becomes any more French between Calais and Paris. Similarly, in *The Mapping of America*, Seymour Schwartz and Ralph Ehrenberg suggest that the series of maps their book collects 'reveals a metamorphosis from artistry to science' (Schwartz and Ehrenberg 1980: 9). Like Taylor's, this survey makes the metamorphosis from art to science seem as immediate and absolute as the transformation from caterpillar to butterfly.

It is not only the modern map which emerges blinking in Schwartz and Ehrenberg's account into the bright light of historical day. As well as chronicling their own 'metamorphosis', the American maps they collect provide 'a graphic picture of the nation and an illustrated chronicle of its evolution' (Schwartz and Ehrenberg 1980: 9). This is a double teleology, doubly paradoxical in its fusion of temporal process and punctual transformation. On the one hand the successive developments of the map mirror the successive stages of an American 'evolution'. On the other each map pictures the same Platonic 'nation'. The butterfly of America and the butterfly of modern cartography emerge together and in symbiosis, each identity as absolute and irreversible as the other.

The first revolution in a critical history of cartography involved what might be called a thickening of the map; a shift from the essentially idealist habit of seeing through it, as a window on the world, to one of reading it as cultural text. This revolution began in the history of art.

The iconological tradition in art history treats the newly geometric spatiality of early modern perspective paintings not as 'a definitive victory over Medieval parochialism and superstition' but as a form of culturally specific, symbolic meaning which can be read (Edgerton 1975: 9). In Erwin Panofsky's seminal thesis, perspectival space is no absolute category of the world but the third, most abstract, 'iconological' layer of three layers of pictorial signification (Panofsky 1955). Whilst Panofsky subscribed to the Kantian notion of mental absolutes, including absolute space, he introduced a mediating cultural gap between such absolutes and their representation by particular cultures. For Panofsky the space of Renaissance perspective painting did not simply represent a newly discovered object world, but expressed distinctly cultural preferences for the way in which this world should be represented. In recent years this Panofskian, iconological approach to art has informed a revisionist history of cartography which seeks to appreciate the geometric map as cultural text. Yet in the main this new history has acted only as a mournful counterpoint to traditional map history, reinforcing its idealism through a set of complementary assumptions about the modernity of the early modern map, and shadowing the traditional view of enlightened cartographic 'discipline'.

Maps, knowledge and power

If it draws upon Panofskian iconology in its preparedness to read the map, this new cartographic history also owes much to French post-structuralism, and specifically to Michel Foucault, in its account of what the map might say. Foucault's analysis of the uses and the politics of Enlightenment space is notoriously pessimistic, treating modern modes of spatial thought and planning as inextricable from the exercise of power. Linnaean botany and the sciences of madness and penalogy map out common ground for Foucault in a 'spatialisation of knowledge' working to define and subject nature and humanity and embodied in the pun of 'discipline' (Foucault 1986b: 254). In both of its senses 'discipline', notes Foucault, 'fixes; it arrests or regulates movements; it clears up confusion ... it establishes calculated distributions' (Foucault 1986a: 208–9). Foucault's descriptions of the 'spatialisation of knowledge' and of the disciplinary uses of spatial planning and representation have proved vastly fertile in revisionist studies of cartographic history, and most prominently in the work of Brian Harley.

One of the most widely read of Harley's essays is included in Denis Cosgrove and Stephen Daniels's 1988 collection *The Iconography of Landscape* (Harley 1988). Harley's essay in this collection reads cartography as a Panofskian 'cultural image', and his title – 'Maps, knowledge and power' – makes clear the Foucauldian parameters within which this reading is performed (Cosgrove and Daniels 1988: 1). Harley regrets that cartographic history has been dominated to date by what he deems a 'positivist' teleology of evolving accuracy. What this history elides, he suggests, is the partiality of modern maps as simply one kind of 'way of conceiving, articulating, and structuring the human world' (Harley 1988: 278). In fact, Harley argues, the 'Euclidean syntax' privileged in post fifteenth-century cartography did not just reflect the world, but 'structured European territorial control' (ibid.: 282). The particular rhetoric of 'authority' explicit in the Medieval map had not gone away, but was now hidden by this 'silent' geometric syntax.

Harley's interventions in this and other essays laid out the ground for a critical history of cartography which examined the role of maps in the cultural making of knowledge. Harley comments, towards the end of his essay, that his ideas 'remain to be explored in specific historical contexts' (Harley

1988: 303). A great deal of work has now been done to answer this call for specific, contextualized explorations of the cartographic power/knowledge nexus. Much of this work has investigated the role of developing cartographic practices in early modern English/British imperial ventures in Ireland and America and nearly all of it has tended to stress the silent 'syntax' of power encoded in Euclidean geometry. In a 1988 article titled 'Inventing America', for instance, William Boelhower depicts the successive phases of an increasing cartographic mathematization and abstraction not as an evolution towards enlightened accuracy, but as responses to the needs of successive phases in the chronology of imperialism (Boelhower 1988: 480). As the final phase in this honing of imperialist tools, Boelhower characterizes the American 'scale map' as a 'panopticon': that epitome of the totalizing spatial logic of Michel Foucault's disciplinary society (ibid.: 496). This narrative is the direct obverse of Schwartz and Ehrenberg's celebratory national/cartographic dawn-chasing. Boelhower sees 'the line's regime cast[ing] its geometrical scheme over more and more of the new continent' like an ominous shadow, 'fixing', and thereby obliterating, native movement and locality in its wake (ibid.: 488).

From product to process

Notwithstanding the pervasiveness of Harley's simultaneously Foucauldian and iconological approach, several notes of warning have been sounded in recent years which have worked to undermine it. These warnings have come primarily from two directions: one, that of traditional, 'positivist' map scholarship; the other, an alternative fork to Harley's in the path of a newly cultural history of geography.

In his introduction to the posthumous collection of Harley's essays *The New Nature of Maps* (2001), map historian J.H. Andrews poses the following rhetorical question:

> Positivist historians have plenty to do when confronted with a previously unknown map. Besides establishing its date and authorship, they can analyze material form, method of drawing or reproduction, use of inks or paints, projection, linework, extent of generalization, choice of symbols, stylistic affinities, sources of information, method of survey, influence on other maps, archival history, distribution, and use. What can the non positivist scholar do except say, 'Just as I thought: more glorification of state power'.
>
> (Andrews 2001: 31–2)

The problem as Andrews sees it is that whilst it is easy for even the lay map reader to decode simple layers of cartographic meaning – inductively, or by using a key – 'there is nowhere they can go to verify the presence of the abstract ideas allegedly embodied in the map' (Andrews 2001: 11). In the absence of any specifically cartographic evidence for the 'abstract' meanings of maps, Andrews finds Harley relying on 'an analogy with other art forms whose practitioners have been more communicative' (ibid.: 11). And this strategy of analogizing leads him to consider not only what Andrews considers inadmissible evidence from other disciplines – 'art history, literary criticism, architecture, and music' – but also 'non-cartographic' elements of maps themselves, including 'decorative embellishments' (ibid.: 11).

Andrews's critique suggests that Harley's iconology over-reads the map, reifying and totalizing its meaning; filling its apparent silences with misplaced rhetorics from elsewhere. Similar warnings have been sounded from a rather different direction. Since Foucauldian New Historicism became conspicuous as a movement in literary studies, a host of materialist cultural critiques have focused on the way in which this approach can seem to further the work of the representational practices it describes, perfecting their forms and re-incorporating that which escapes them as part of the 'system' (see, for instance, Burt and Archer 1994). Responsibility for this theoretical totalizing can be traced directly to Foucault: first for the closure which he attributes to the modern 'disciplinary society' and its 'indefinitely generalizable mechanism of panopticism', and second for the formalism by which he models it (Foucault 1986a: 206). Foucault himself acknowledged – in dialogue with a group of geographers – his use of an analytic lexicon replete with unexamined spatial metaphors: of 'implantation, delimitation and demarcation ... the organisation of domains' (Foucault 1980: 72).

New cartographic historicists, to coin a rather awkward label for such Foucauldian map-readers as Brian Harley and William Boelhower, can seem highly vulnerable to this materialist critique. Rather than relating representations to their specific local conditions of meaning and use they often map formal patterns discovered in their texts onto spatialities still more abstract and idealist than those of Enlightenment geometry. Moreover they can often seem to elide the gap between these aesthetic and conceptual spatialities and the space of practical activity, as if the map really were an encapsulation of the world. In his analysis of American cartography, for example, Boelhower projects a battle between an imperialist geometry which seems to have its own agency, and resistant cartographic toponyms whose inherent particularity opens 'a trap door ... in the written surface of the map' (Boelhower 1988: 494). Julia Lupton writes of rebel resistance to English cartography in Ireland 'cracking, piercing and mutating' the colonial geometric plane (Lupton 1993: 93).

From the materialist point of view these idealist slippages between abstract space and the space of human interaction are the product of a characteristically Foucauldian over-estimation of representation itself. Attacking the dominant language model of cultural analysis and demanding a Marxian critical shift from products to processes, Henri Lefebvre has complained of Foucault that he:

never explains what space it is that he is referring to, nor how it bridges the gap between the theoretical (epistemological) realm and the practical one, between mental and social, between the space of the philosophers and the space of people who deal with material things.

(Lefebvre 1991: 4)

Panofsky encouraged the student of art to read widely in order to historicize their intuitive interpretation of artistic symbolism. In his monumental history and critique of spatial production Lefebvre constantly urges caution in this critical turn to language. Real space for Lefebvre is social space, and it is produced through processes in which the abstractions of verbal media play no especially privileged role. Why, he asks, should language be granted the special status Foucault and his ilk implicitly accord it? 'Does language ... precede, accompany or follow social space? Is it a precondition of social space or merely a formulation of it?' (Lefebvre 1991: 16).

Materialist cultural geographies have often accorded with these warnings about language and representation. Peter Jackson, a geographer highly instrumental in importing cultural studies methodology into his discipline, insists that his call for a 'more expansive view of culture' shouldn't lead to the over-privileging of linguistic cultural forms (Jackson 1989: 7–8, 9). In writing published since *The Iconography of Landscape* Denis Cosgrove has worried about criticism preoccupied with reading the disciplinary aesthetics of the map itself (Cosgrove 1999: 1). When we shift our focus from product to process, he suggests, we soon see the 'aesthetics of closure and finality dissolve' (ibid.: 2).

These materialist critiques suggest that we should tread very carefully indeed before reading maps as 'cultural images', rather than local interventions in material social processes. Yet notwithstanding the warning signs staked out along disciplinary boundaries, a new cultural history of geography has placed considerable emphasis on writing and has pushed the analogy of reading far beyond the bounds of written texts. The 'interface', as one scholar calls it, between literary and cartographic study is proving massively fertile ground, and yet what or where exactly is this interface (C.D. Pocock 1988)?

Idealist analogies and the post-structuralist critique

Much of the most recent work on the relationship between geography and literature seeks to establish a broader context for spatial representation than that of local processes and transactions.[2] At its most challenging this kind of work relates literature and cartography in terms of overlapping modes of cultural production, subject to distinct but related social pressures, mediated by distinct but related generic codes. In its attention to the local limitations both of social process and of generic form it is able to ward off much of the cultural formalism of which New Historicism stands accused.

Less secure, however, are those many instances in which readings of literature with cartography still make formal comparisons over an abstract and idealizing common ground, 'reducing' the cultural specificity of their subjects. Most literary scholars of cartography can be accused to some degree of that preoccupation with the aesthetic associated with Foucauldian New Historicism. Moreover, rather than re-integrating the formal abstractions of cartography and literature with the local processes of production and consumption which generated them, these readings often reinforce them through idealizing analyses of the 'space' engendered by cartography and literature.

In their seminal work on literature and cartography Richard Helgerson and John Gillies both give considerable weight to the formal correspondences between maps and literary texts (Helgerson 1992; Gillies 1994). In Helgerson's case these formal resemblances are mapped onto the more general common ground of cultural pressures, and specifically onto an ascending whiggism calling for a decentralized national spatial imaginary. In John Gillies's case, however, the general common ground underpinning particular, aesthetic resemblances between maps and literary works is shaped not just by contemporary political consciousness, but also by subconscious human impulses to stratify and thereby textualize space, marking the scene, the obscene, and so on.

The most common relationship between maps and verbal texts discovered in recent scholarship is similarly formal and phenomenological. Most critics reach for a mobile, metaphorical definition of what maps and literary texts are and do which will accommodate and permit comparison. Both Tom Conley and Rhonda Lemke Sanford, for instance, identify early modern literary works which seek, like conventional cartography, 'to contain and appropriate the world they are producing in discourse and space through conscious labours of verbal navigation' (Conley 1996: 5, cited in Sanford 2002: 13). Bernhard Klein, in turn, has argued that both literary and cartographic texts can be categorized as either static map or mobile itinerary, depending on the relationship they establish between reader and space (Klein 2001).

The danger in this tendency to read maps and literary texts in terms of spatial symbolism or analogy is that it formalizes in advance our view of particular social processes. It assumes that spaces 'framed' by geometry or 'navigated' in verse felt to early modern readers much as they feel to us, and thereby naturalizes the advent of 'modern' forms of spatiality and representation, however much it may appear to mourn them. It cleans up the dirtiness of the early modern map. But where positivist and materialist critiques have blamed an excessive post-structuralism for these abstractions, a final, and I think most telling, critique of the new cartographic historicism blames an insufficiency.

In his critical introduction to Brian Harley's essays, self-confessed positivist J.H. Andrews finds Harley asking his reader to question the 'assumed link between reality and representation' but notes with relief that Harley draws back from the post-structuralist brink of finding nothing 'outside the text' (Andrews 2001: 21). Barbara Belyea, on the other hand, finds this hesitancy problematic (Belyea 1992). To bring ornament to the centre of the map is to accept that maps, like other texts, do indeed – in Andrews's incredulous phrase – 'create noncartographic reality as well as representing it' (Andrews 2001: 11). And yet, Belyea points out, Harley's work on cartography consistently supposes a normative physical reality, politics, ethics and human subjectivity which cartography distorts and represses (Belyea 1992: 4).[3] Whilst this supposition is in perfect harmony with the idealist, Kantian basis of Panofsky's approach, it does not sit well with the post-structuralism with which Harley tries to mix his iconology (ibid.: 2).

For Foucault and Derrida, Belyea observes, political power is not external to the text, and executed upon or through it, but is inextricable from and a product of textuality and discourse (Belyea 1992: 3). Truth is not something which human subjects misrepresent and suppress through textuality and discourse, as Harley suggests in his readings of cartographic 'silence', but is a product of textuality and discourse themselves. As, for that matter, is the human subject. Maps do not simply 'hide' power in those margins which positivist scholars would have us believe are not part of cartography. Rather, they make it possible precisely in their marking of the boundary between centre and margins; truth and ornament; representation and reality.

Belyea's critique suggests that the characteristic slippage between practical and aesthetic that we find in Harleyan readings of cartography is the product not of an over-estimation of representation and language, as materialists have suggested, but of a half-hearted post-structuralism which sees representation as the 'tool' of political agencies operating somehow beyond it. I would apply the same critique to recent readings of cartography and literature which have sought in phenomenology a refuge from Foucauldian pessimism and an idealized common ground beyond cartographic discipline.

For the phenomenological tradition in philosophy there is no possibility of Cartesian detachment and the *'cogito ergo sum'* (Crang 1998: 107). No possibility, that is, of a subject that might regard the world objectively and separately. Being, as Martin Heidegger put it, is always *'dwelling'*, or *'being-in-the-world'*, and the self, rather than being limited by physical boundaries separating it from the world, is constituted through such boundaries (ibid.: 107). Phenomenology appeals for us to examine images, whether visual or literary, not as substitutes for an objective 'reality', but as the way in which we experience our world.

Whilst phenomenology is indebted to Kant for its sense of the mental mediation of space, it rejects the Kantian notion of space as an absolute category even of the mind. Space, to use Edmund Husserl's language, is 'intentional'; or to use Heidegger's, imbued with 'care': it is constituted and shot through with human negotiations, processes and desires (Crang 1998: 108–10). The best the philosopher can do is search, as Husserl does for geometry, for the essence of the human experience of a phenomenon: the sense it must have had for its first discoverers, with all the intervening overlay of history bracketed or reduced.

For all its insistence on an essential subjectivity and humanity, this quest for the heart of the phenomenon remains an idealist one: an attempt to re-ground knowledge on something absolute and eternal. In recent years it has been subjected to a persistent post-structuralist critique, most prominently in the work of Derrida. Derrida brought this critique to bear specifically on Husserl's attempt to re-ground geometry on realities beyond history and language. I think it applies with equal force to the attempts made by recent literary critics to read in textual and cartographic images the traces of archetypal human experiences of space.

Husserl insists – 'obstinately', in Derrida's view – that the objectivity typified in geometry lies behind, and is the condition of possibility for, language and history itself (Derrida 1978: 61–6). This insistence begs an archetypal Derridean question: if geometry is prior to language and history, and yet not absolutely ideal, why and how was it invented, and by what means might the pure sense of this invention be experienced and transcribed? Since language and history are the only media for either moment of invention, this question has no satisfactory answer.

Recent modifications of the phenomenological tradition have looked to qualify its essentialism, but do not dodge this Derridean critique. Some scholarship has tried to resolve the tension between phenomenology and history by emphasizing the 'embodied' nature of human experience: its inextricable enmeshment not only with the human body and with worldly phenomena, but with specific material fields of history, locality and culture (Crang 2000: 19). Yet still the phenomenological category comes first and lies behind history, culture, and any particular embodiment in representation, begging the Derridean question: if culture mediates the phenomenon, what is to stop it altering or even reversing its 'true' identity? Along with notions of embodiment and acculturation, an increasing emphasis on mobility has also sought to revise and rescue the idealism of that stable relationship found in traditional phenomenology between self and place (ibid.: 19). Michel de Certeau, for instance, invokes a notion of performance, analogous to the everyday use of language. 'The act of walking', he suggests:

> is to the urban system what the speech act is to language or to the statements uttered … it is a process of *appropriation* of the topographical system on the part of the pedestrian (just as the speaker appropriates and takes on the language); it is a spatial acting-out of the place.
> (de Certeau 1984: 97–8)

Such performances work for de Certeau to reverse a process of cartographic alienation. The abstract, mathematical map, he suggests, 'has slowly disengaged itself' from the stories and itineraries that were the 'condition of its possibility' (de Certeau 1984: 120). Spatial performances involve a return from the 'artifice' of abstract space to the nature of absolute space, owned by its users. Once again, however, post-structuralism makes these disinctions and analogies between language, geometry and practice seem highly problematic.

To suggest that the abstract mathematical space we find in early modern maps was imposed in the real social world against a resistance somehow absolute, a-historical and beyond language and representation, is to reduce the contingency of language itself as the medium through which new modes of building and representing space were established and contested, and to ascribe to it precisely the same monumentality for which Lefebvre blames post-structuralism. In fact, Derrida's own question about the relationship between geometry and language matches and answers the Lefebvrean one. Language is neither prior to the human experience and representation of space nor posterior to it. Writing, geometry and practice are not analogous but inextricable and the same. A thoroughly post-stucturalist critique of the new critical history of cartography suggests that its revolution has been incomplete. But it does not accept that we are wrong to 'read' when we do cartographic history; simply that we are wrong to read the map as an analogy or alternative to language; and especially wrong to attempt intuitive readings of geometry and space.

We cannot, as Husserl hoped, share the experiences of early modern subjects by imagining the geometries and other spatialities encoded in their texts. We cannot do this because these experiences are not extricable from history and language. And when we abandon this Husserlian quest and appreciate the written-ness of early modern geometries and geographies we find, in fact, that they were far from being what they seem intuitively to us. In reality, the currency of geometry in the seventeenth century, and accordingly of the geographies it informed, was highly unstable. Far from being the symbolic form through which early modern subjects inevitably perceived their worlds, far from being the 'silent' ground, the naturalized basis for a 'disciplined' experience of space, the meaning of geometry and the map was contingent on a cacophony of rhetorics contesting and negotiating their interpretation. As literary historians have suggested in analyses of those early modern meta-narratives that accompanied the birth of the novel, these rhetorics were often simultaneously rhetorics of 'truth' – of the right way to represent – and rhetorics of 'virtue' – of the legitimacy of those individuals and communities who produced and were figured in early modern maps (McKeon 1987).[4] In these meta-narratives of truth and virtue we can begin to discover a genuinely cultural relationship between writing and cartography.

Between the study and the marketplace

J.H. Andrews may be right, at least for the eighteenth and nineteenth centuries, when he judges modern cartography an intrinsically silent, non-rhetorical art: 'before about 1930, cartographers made few general pronouncements of any kind about their subject' (Andrews 2001: 5). But for a substantial part of the sixteenth and seventeenth centuries, cartography was noisily rhetorical, signifying not just 'outwards' to the spatial world, but 'sideways' to discourses of legitimate and truthful representation. Sixteenth- and seventeenth-century mathematicians and geographers worked hard to establish the parameters of truth and virtue within which their knowledge and work would be understood. Sometimes they wrote directly to preface a particular map or set of maps, as the cartographer John Norden did in 1596 (Norden 1596). In other instances they wrote more generally to define and advance the currency of mathematics and geography. The best-known aspect of such advertisements has fuelled the idealist conception of a clean mathematical panopticism. But it is only one side of the story.

Geography, claims mathematician and physician William Cuningham in a much-cited passage from *The Cosmographical Glasse* (1559), 'delivereth us from greate and continuall travailes. For in a pleasaunte house, or warme study, she sheweth us the hole face of all th' Earthe, withal the corners of the same' (Cuningham 1559: sig.A6r). Dedicated to Elizabeth I's favourite, Robert Dudley, Cuningham's treatise promises to teach its reader how to draw a map for 'Spaine, Fraunce, Germany, Italye, Graece, or any perticuler region: yea, in a warme and pleasaunt house, without any perill of the raging Seas: danger of enemies: losse of time: spending of substaunce: wearines of body, or anguishe of minde' (ibid.: 120). Cuningham's offer to place his patron and his reader above the world; beyond travail; outside even their vulnerable bodies represents by far the best known aspect of the early modern 'cartographic transaction' (Klein 2001). It is a manifestation of that 'Euclidean ecstasy' which infused early modern scientific culture from the sixteenth century onward, inspiring aspirations for a new dominion over nature and humanity (Cosgrove 1993). Yet equally common in mathematical writing are figures and rhetorics which dirty somewhat the clean lines of geometric discipline.

Cuningham also tells of maps which Alexander, 'the mighty Conqueroure', would have made of the country 'with which he would warre', and would have 'hanged in open markets for all men to behold, wherby the Capitaines did forsee, and seke out where was the easiest places to arrive, and the Souldiors allured with the commodities of the Countries, were made the willinger to the thinge' (Cuningham 1559: sig.A4r). These maps take us far from the scholar's study and into a world of political strategy, commercial commodity, material pain and pleasure. Moreover, alongside polar images of detachment and engagement, Cuningham presents images which equivocate between the two.

In an account derived from Ovid's *Metamorphoses* Cuningham tells of Daedalus 'that excellent Geometrician', who saw the 'Monster Ignorance' with 'the eyes of knowledge' and, with wings prepared '(throughe Science aide)', flew 'oute of hir mooste filthy Prison'; 'her lothsome Labyrinthe'; 'Ascending to the Sterrye Skie' (Cuningham 1559: sig.A2r). Knowledge, concludes Cuningham, shuns ignorance; brings man closer to God; and permits the invention of arts through which man has 'sought out' worldly 'Secretes' (ibid.: sig.A2r). Here we appear to be back in the scholar's study. But we and every imaginable contemporary reader of Cuningham's treatise know two things that complicate this story: that Daedalus's son Icarus paid a terrible price for starry soaring in the flight from Crete, and that the labyrinth from which the pair escape was built by Daedalus himself. Science, it appears, is both escape route and trap; both of the world and out of it. In Francis Bacon's words, 'The *Labyrinth* is an excellent Allegory, whereby is shadowed the nature of Mechanicall sciences ... for Mechanicall arts are of ambiguous vse, seruing as well for hurt as for remedy, and they haue in a manner power both to loose and bind themselues' (Bacon 1619: 93–4).

The culture of discipline

In his interpretation of the labyrinth myth Francis Bacon expresses an ambivalent view of the artful application of mathematics: one which sits uncomfortably with his place in the intellectual history of the seventeenth century, and yet which I think aptly characterises the unstable currency of early modern science.

Bacon is famous for his articulation of a radical humanism which shifted the attention of science from idealist contemplation to the improvement of the human condition, and which reduced the material world to 'disciplined', manipulable mathematical form. The Baconian philosophy of discipline and improvement is widely viewed as the ethos informing both the mathematization of seventeenth-century geography, and the economic reformism and territorial expansionism for which this newly mathematical perspective is held to have served as instrument and ideology. Yet this imbrication of capitalism, imperialism and Baconian science hides significant tensions in seventeenth-century English culture: tensions manifest in Bacon's reading of the labyrinth.

Even the most pragmatic, worldly streams of Protestant humanism, equating truth and virtue with utility, contended throughout the seventeenth century with a residual discourse of Calvinist asceticism, associating practical art and economic development with moral and social corruption. In the light of this sustained cultural ambivalence, it should be unsurprising that much seventeenth-century geography conveys a mixed message about what is virtuous in the use of land, and what is true in representing it. The art of mathematical surveying is often the centrepiece for cultural histories of the commodification and colonization of early modern 'space'. Yet for every passage in a seventeenth-century surveying manual which promotes fantasies of mathematical reduction and discipline there is another which insists upon the legal and social complexity of customary landscapes and of the surveyor's work in representing and maintaining them. Rather than cleanly turning new social values into new disciplinary truths, the geographic practices of seventeenth-century England were characteristically caught between the verbal transactions of manorial stewardship and the mathematical reformation of land.

Moreover, not only was seventeenth-century geography less comprehensively mathematical, less disciplined than we might think, but the role of mathematics itself in these equivocal, transitional geographies was not what we might expect it to be. Rather than silently marking out the new territory of capitalist and imperialist spatial discipline, I think mathematics came in seventeenth-century cartography to play something more like the moral role of traditional feudal stewardship itself, forming part of a rhetoric of balance and constraint; of arbitration between liberal virtue and worldly profit.

Middle men

Early modern mathematicians and geographers who published to promote their knowledge and their arts typically hedged their bets, like William Cuningham, between liberal scholarship and profit; between the study and the marketplace. I think this textually fashioned rhetoric of intermediacy did two kinds of cultural work. It served the Protestant-humanist and Puritan cultures of sixteenth- and seventeenth-century England as a means of negotiating the problematic status of artfulness, profit and the world. But it also allowed mathematicians and geographers themselves room for manoeuvre: for self-fashioning.

The Harleyan, New Historicist account of early modern cartography sees it as clean and disciplinary; intolerant of the slightest departure from impersonal, mathematical authority. Yet I think seventeenth-century geographers make these departures remarkably conspicuous. Many mathematicians and geographers draw attention to the painfulness, uncertainty and danger of their work, telling stories of being lost in impenetrable wildernesses; of geographies performed in the heat of battle; and of negotiating with 'savage' informants without whose help they could not have made their maps. In the language of classroom mathematics they present not just solutions, but also the 'work' it cost to produce them. They construct themselves as mediators between their readers, patrons or clients and a dangerous, doubtful, sinful world. And they remind us that a mathematics which truly conformed to the Husserlian ideal, with all its historical and cultural residues reduced, would be as empty and as meaningless as the disciplined spaces of Brian Harley's maps. To have currency, agency and meaning, as both Bacon and Derrida have recognized, mathematics must be wrapped in the binding, loosing labyrinth of language.

The remainder of this book can be divided into two parts, addressing my two key contentions. The first part, comprising Chapters 2 to 5, is concerned to re-think the role of mathematics and geography within a Protestant culture of discipline. It will argue that they served more as sandy rhetorics of negotiation between traditional and new truths and values than as clean ideological and practical instruments of reform. The second part of my book, comprising Chapters 6 and 7, explores the ways in which seventeenth-century self-fashioners exploited the unstable currency of mathematics and geography in an environment of scepticism and conservatism, fashioning flexible social roles and agencies for themselves as mediators between theory and practice; the study and the marketplace.

Two final words on writing, geometry and space. First, the reader may note a persistent slippage in my terms of reference between geometry and mathematics; between geometry and geography; and between various genres of early modern geography. These slippages are tolerated because I'm more concerned with the reputation or cultural currency of these various entities than with making what the modern reader would recognize as correct distinctions between them. In most cases the values attached by early modern writers to one branch of the mathematical arts applied equally to another branch, or to the roots themselves. Second, in a book which deals extensively with maps, it might surprise the reader that there are only three illustrations that look remotely cartographic. The illustrations I include here tend to be examples less of spatial representation than of discourses on spatial representation. This is a book which makes little attempt to read maps 'themselves', dealing instead with their cultural currency in textual discourse. As the preceding pages have demonstrated, I think early modern geometry and geography were radically rhetorical and remain meaningless beyond language. They cannot, in themselves, be read.

2 Discipline and polish

The age of improvement and its limits

Behind the widespread conception of early modern geometry and geography as 'discipline' lies a broad consensus in the recent cultural history of the English seventeenth century which casts the period as an age of individualism and economic reform. The following chapter will explore this consensus and introduce my challenge to it.

I begin with what I take to be some received ideas. The English colonization of America and the capitalization of English agriculture generated and were advanced by a labour theory of economic value, expressed definitively by John Locke, but ascendant throughout the seventeenth century. This theory licensed the appropriation of 'inefficiently' used indigenous and common land in the interests of 'improvement'. It was intimately connected in an imperializing and capitalizing culture with a humanist theory of scientific value, expressed definitively by Francis Bacon, which shifted the origins and ends of science from idealist contemplation to the improvement of the human condition, and which subjected the material world to mathematical abstraction and manipulation: to 'discipline'. These seeds of change were nurtured by the Puritan culture of early modern England, which sought the truth of science in the benefits it yielded to the Godly individual and commonwealth, and which set discipline and industrious reform against indolent custom. Their green shoots intertwined in arts that developed to secure and celebrate the providential progress of the elect individual and nation, demonstrating their election in their industrious success. Amongst these arts geography in its various forms was signal. Geographies of England and its colonies in the seventeenth century expressed the values of individuals who discounted any form of land use beyond productive private property; cultivated science in the interests of the industrious and Godly; and took geometry for granted as the measure of the disciplined material world. In short, the new, mathematical geographies of the seventeenth century heralded what one historian, invoking the watchword of the early modern capitalist, has dubbed the 'age of the improver' (Darby 1973).

A short history of improvement

The dawn, in cultural histories of the seventeenth century, of the 'age of the improver', marks the dusk of a Medieval view of land and society. Until at least the sixteenth century, the political identity of England was profoundly invested in the notion of a moralized agrarian economy. This moral economy was envisioned in terms of a descending hierarchy of stewardships connecting King to commoner, each steward responsible for the customary rights and responsibilities of the estates beneath him. Land within this 'moral economy' was considered not as absolute, private property, but as held in trust for the wider community it maintained: hence the central importance of 'stewardship'.

The gradual capitalization of English agriculture, beginning with the pasture enclosures of the Tudor squires, presented a singularly important challenge to these traditional values, and until the seventeenth century enclosure was countered consistently both by restrictive legislation and a tide of opprobrium from pen and pulpit. Yet by the mid-seventeenth century enclosure had begun for the first time to win cultural legitimacy and legislative support.[1] The challenge for cultural history has been to explain the role of language and representation in negotiating this shift in values.

The most significant recent scholarship in this area has come to place an ever-greater emphasis on a rising ethos of individual endeavour, expressed mutedly in literary appropriations of classical Georgic, and more stridently in explicit rhetorics of economic reform. Andrew McRae has described the development of agrarian discourse from the moral economy of the Tudor Protestant complaint tradition which condemns all change as corrupting covetousness to the gradual emergence of a discourse of 'Godly individualism' valorizing thrift and embracing profit (McRae 1996). In mid seventeenth-century treatises promoting improvement not just to the landlord but to a wider audience, and in 'project'-oriented tracts associated with such speculative ventures as fen drainage, he traces the emergence of a 'discourse of improvement' which explicitly embraces industrious individualism, challenging the traditional moral economy (see, for instance, Blith 1649 and 1652; Sir Jonas Moore 1685). These texts present artful husbandry not just as the responsibility of the industrious patriarch, but as the basis for a universal improvement of the nation (Figure 2.1).

For McRae the revolutionary significance of a rising discourse of 'improvement' turns on the shifting cultural meaning and value of improvement itself. McRae suggests that the discourse of improvement unites two originally distinct terms: 'approve' (to appropriate) and 'improve' (to husband artfully), bringing the positive values associated with artful husbandry to bear on the contentious act of appropriation (McRae 1996: 136). This explanation underestimates a little the problematic and shifting status of art in the seventeenth century: the crux of the present book.

In fact, 'improvement' had always had the two meanings McRae

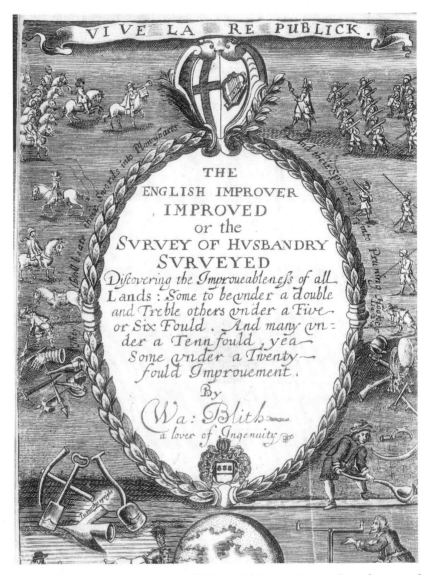

Figure 2.1 Walter Blith: frontispiece of *The English Improver Improved; or, the survey of husbandry surveyed* (London, 1653). Reproduction courtesy of the British Library (shelfmark: 234.e.44).

describes: of appropriation and of artful husbandry: these meanings ran in parallel (OED 'improve'; OED 'approve'). Rather than the latter displacing the former, what changed, to some degree at least, was the value attached to productive art itself: to sciences of the world. This change, which subsequent chapters of the present book will insist was partial, gradual and constrained, can be attributed to the largely congruent influences of Protestant humanism, seventeenth-century Puritanism and Baconian science.

The Florentine Renaissance, as embodied in the translations and commentaries emanating from the Academy of Marsilio Ficino and Pico della Mirandola, retained a classical Hellenic and Medieval bias towards contemplative scholarship. Yet as the philosophy of Renaissance humanism developed it also placed a growing emphasis on the Aristotelian moral value of civic participation (J.G.A. Pocock 1975: 58). Accordingly, alongside its commitment to a contemplative metaphysics, Renaissance humanism cultivated more worldly arts of government, painting, architecture and rhetoric. In rhetoric – the study not just of ideas, but of their means of expression, pursued through conversation with ancient authors – Renaissance humanism expressed its belief in human association, in utility, and in the historical enactment of abstract truth and virtue (ibid.: 59–63).

The sixteenth-century reforms in English education set in motion by Erasmus, Vives and Ascham leant heavily towards this civic, Aristotelian aspect of contemporary humanism, and promoted literary disciplines modelled on revived Latin texts. Reformation humanists showed antipathy towards contemplative, theoretical science as both un-civic and offensive to the Calvinist conception of a supreme divine will before which human reason must be humble (Feingold 1984: 29). Commissioners sent by Edward VI to Oxford in 1550 burnt books containing mathematical diagrams (Yates 1964: 166). Such 'polite learning' as the English humanists did countenance was of necessity 'moderate' and, as Ascham put it, geared to 'som good use of life' (quoted in Feingold 1984: 30; see also Katherine Hill 1998: 260–1). The English mathematician Thomas Digges strikes a characteristic contemporary note when he describes in a treatise of 1579 his turn from esoteric to practical, sociable and public mathematical studies:

> The more subtile part of these Mathematical Demonstrations did breede in me for a time a singular delectation, yet finding none, or very few, with whome to conferre & communicate those my delites, (& remembring also that graue sentence of Diuine Plato, that we are not borne for our selues, but also for our Parents, Countrie, and Friends.) After I grew to years of riper iudgement, I haue whaley bent my self to reduce those Imaginatiue Contemplations, to sensible Practicall Conclusions: as well thereby to haue some companions of those my delectable studies, as also to be able, when time is, to employ them to the seruice of my Prince and Countrie.
>
> (Digges and Digges 1579: sig.A3r–A3v)

By the turn of the seventeenth century, writies P.M. Rattansi, and Bacon's projection of a great reformation of learning, 'the idea of studying Nature to glorify God and benefit mankind through inventions and discoveries was something of a commonplace' (Rattansi 1972: 12).

Commonplace or not, in Bacon's writing the Protestant-humanist philosophy of scientific utility achieved a new clarity and schematic rigour. Bacon believed that just as religion and faith might in part regain for man the innocence he had lost with the fall, arts and sciences cultivated with a view to use might to some degree regain for him the dominion he had thereby lost over nature. Accordingly he sought to explode the traditional separation of science into theory and practice, urging intellectuals to set aside their hereditary contempt for useful knowledge; to value knowledge only as it contributed to power over nature and the improvement of human life; and to pay attention to individual phenomena rather than abstract models. Bacon belived that only the mechanical arts, so often despised by scholars, continually improved. Theoretical science, detached from the world of things and practices, was apt to become corrupted. In fact the mark of value in the new empirical science was not just its active, but even its violent and laboured quality. Digging down to the roots of knowledge the conquering Baconian scientist exhibited not arrogance before nature, but humility, seeking sympathy with and thereby power over its individual qualities.

The impact of Bacon's philosophy of science was contingent on an intensification of the Protestant values that underscored it. In its promotion of practical science as the means to universal human improvement and a worldly paradise regained, Bacon's approach meshed with the radicalizing Protestantism of the seventeenth century, which increasingly turned from general truths rendered meaningless by the absolute will of God to the 'relative, finite ends' left open to human life (Cassirer 1970: 70). Whilst the Baconian ideals of scientific philosophy and method gained little purchase during Bacon's lifetime, they began to achieve serious prominence in the mid seventeenth century, appealing to an increasingly dominant Puritanism and to the social and educational reformism of the Long Parliament (Rattansi 1972: 27). Forming, in Charles Webster's words, almost 'the official philosophy of the revolution', these ideals were urged in an outpouring of plans for technical innovation and social, scientific and educational reform (Webster 1975: 25; Webster 1979: 37–8).

Much of this outpouring was associated with the circle gathered around the German-born archivist of scientific knowledge Samuel Hartlib, who lived in England from 1628 until his death in 1662, and formed the lynchpin of an intellectual salon hosted by Robert Boyle's sister Lady Ranelagh (Rattansi 1972: 18–19). As well as showering Parliament with reformist proposals, Hartlib spread his ideas amongst a wider audience through his own writing and through collecting, coordinating and publishing the writing of his associates. The Hartlib circle produced a clutch of proposals circulated privately and publicly which attended simultaneously to the spiritual and physical condition of mankind: texts in which McRae and others have found the purest strains of a rising seventeenth-century 'discourse of improvement'. Gabriel Plattes, for instance, wrote both *The Profitable Intelligencer* (1644), a husbandry manual designed to help its reader unlock the 'infinite treasure' hidden within nature, and *Macaria* (1641), a utopian tract whose social design incorporated a 'councell of husbandry' making improvement first principle of land tenure (Plattes 1979: 4). In a letter to Hartlib in 1644 John Milton set out a 'design' for the reform of English education which struck much the same notes, recommending classes in husbandry to inspire the improvement of the nation's wasted lands and help 'repair the ruins of our first parents' ('Of Education', Milton 1953–82, II: 366–7, 389).

In these aspirations for dominion through improvement cultural historians have sensed the drive of impulses which strained against the customary status quo, and even against the territorial limits of English sovereignty. Indeed the ambitions of Commonwealth reformers were often geographically far-reaching. Alongside its 'Councel of Husbandry' Plattes's *Macaria* incorporated a 'Councell for new Plantations', responsible for sending out the human 'surplusage' of the national economy to found new settlements, 'strongly fortified, and provided for at the publike charge, till such times as they may subsist by their own endeavours' (Plattes 1979: 5). This interest in plantation was widely shared amongst Plattes's peers. Hartlib's correspondents included Massachusetts governor John Winthrop, who sent back to England news of the potential of his American environment, and received in return encouragement and information designed to help him improve it (Haak *et al.* 1878; Webster 1979: 48). When, in 1628, Winthrop founded his 'New England Company for a Plantation in Massachusetts Bay', he was inspired, writes Charles Webster, by a 'Puritan belief in the spiritual value of productive industry' (Webster 1975: 34). Moreover, improvement served Winthrop and others like him not just as inspiration, but as justification for their endeavours.

As long as it has been possible to speak of a general ideology or 'discourse' of American colonization, a theory of property and value based on the 'improvement' of 'waste' land has been its best-known characteristic. Colonial representations of America are notorious amongst historians for their tendency to evacuate their subject: to deem it empty of significant habitation, or at least of legitimate use, and thereby open for appropriation.[2] There is plenty of evidence in the diverse English writing of early America to underwrite these views – both writers who went to America, and armchair propagandists; both Virginia speculators, and New England Puritans drew at times on a rhetoric justifying appropriation through improvement of a vacant or at least neglected land. This rhetoric was designed, according to Allan Kulikoff, to appeal to the aspirations of the middling, yeoman class of small investors who increasingly formed the majority of those who crossed the Atlantic in the seventeenth century, particularly to New England, and whose 'desire for communal rights, familial self-sufficiency, and independence' was thwarted in the domestic economic climate and stoked by the fantasy of free American land (Kulikoff 2000: 17). It was deployed with a particular intensity in New England, where the economic impetus driving a hunger for empty land was reinforced by a heightened sense of religious mission.

What New England Puritans desired above all, according to Richard Slotkin, 'was a *tabula rasa* on which they could inscribe their dream' (Slotkin 1973: 38). Like classical Georgicists embracing the conditions of a world after the Golden Age, they viewed improvement as the duty of a fallen mankind. As Plymouth governor William Bradford puts it in his *History of Plymouth Plantation* (composed *c.*1630–50; first published 1856), exhorting his brethren to 'separate planting': 'Let none objecte this is man's corruption, and nothing to the course itselfe. I answer, seeing all men have this corruption in them, God in his wisdome saw another course fiter for them' (Bradford 1946: 146). Bradford describes the object of Puritan colonial design as 'some of those vast and unpeopled countries of America, which are fruitful and fit for habitation, being devoid of all civil inhabitants, wher ther are only salvage and brutish men which range up and down' (ibid.: 46–7). Puritan commentators involved in and writing to support the migration to New England argued that to enclose and bestow husbandry upon unsettled and uncultivated land was both to respond to a divine command and to earn a right to property (Carroll 1969: 181–2). In a 1629 tract titled 'General Considerations for the Plantation in New England', Bradford's counterpart in Massachusetts, John Winthrop, argued that the Indians had not subdued their land, and therefore, whilst they might have a 'natural' right to it, by God's primary donation to mankind, they lacked the 'civil' right enjoyed by those who 'appropriated ... by enclosing' (Winthrop 1846: 276).

Although his Puritan rhetoric lays bare and purifies the improvement rationale, many historians consider Winthrop's conception of the legitimacy of artful appropriation the general philosophy of English settlement in America. As the seventeenth century progressed, writes Joyce Chaplin, plantations came to be conceived 'within the terms soon to be identified with John Locke, [as] a landscape remarkably transformed by English hands and a demonstration that a civil people could create property out of nature' (Chaplin 2001: 231).

And so, via Protestant humanism; Bacon, Hartlib and the Commonwealth; Winthrop and America; we arrive at Locke, and what most commentators regard as the culmination of a rising ethos of improvement. In *Two Treatises of Government* (composed *c.*1680–2; first published 1689), John Locke (1967) set out a labour theory of ownership and economic value. Locke, like his predecessor in the theorization of natural law, the Dutch Jurist Hugo Grotius, believed that the world had originally been given to mankind in common. Both men concerned themselves with the question of how, in this case, any individual or state might claim a property within this private donation. Both concluded that it was use that defined appropriation, but with an important difference in their understanding of what constituted use. Grotius made a distinction between property in movable objects, whose use and thereby ownership could be constituted through 'attachment', and property in immovables such as land, where use consisted simply of occupancy, generally demonstrated through enclosure (Arneil 1996: 47). Locke, on the other hand, in the fifth chapter of his Second Treatise, titled 'Of Property', found use and thereby a right of property to be constituted universally through the addition of labour, or 'improvement'. For Locke such improvement constituted in itself the enclosure required by Grotius to define immovable property: '*As much Land* as a Man Tills, Plants, Improves, Cultivates, and can use the Product of, so much is his *Property*. He by his labour does, as it were, inclose it from the Common' (Locke 1967: II. 32, pp. 290–1).

Until recently, the most authoritative scholarship on Locke's *Two Treatises* has regarded them as primarily designed to formulate theories of civil society that would engage with the particular circumstances of England in the 1680s and 1690s: the England of the exclusion crisis or of the glorious revolution, according to rival theories of their date of composition. Locke's treatment of America in the Second Treatise, encapsulated in the famous phrase 'in the beginning all the world was *America*', is regarded principally as a figurative means to these English ends, offering an illustration of the original, natural state from which civil society has emerged (Locke 1967: II. 49, p. 301). If a Lockean labour theory of value was drawn into discourses and practices of American colonization, convention states that it was largely *post facto*, or in colonial applications of proto-Lockean theories such as Grotius's. Barbara Arneil has turned this account around, suggesting that the personal involvement of both Grotius and Locke in national and private colonial ventures was a primary motor for their formulation of theories of property and value designed to 'trump' the rights of indigenous inhabitants and rival colonial states. Grotius wrote to serve the interests of the Dutch East India Company, competing with the Spanish and Portuguese over trade in the Eastern Indies (Arneil 1996: 46–7). Locke, suggests Arneil, wrote to serve the interests of his patron Shaftesbury, one of eight lords proprietors of Carolina whom Locke served as chief secretary. Grotius's theory of ownership through occupancy and enclosure was designed to defend his nation's free access to international waters which could not be so occupied or enclosed. Aware, writes Arneil, 'that the aboriginal nations England had encountered in the new world could claim property through Grotius's right of occupancy, Locke developed a theory of agrarian labour which would ... specifically exclude Amerindians from claiming land' (ibid.: 64). In its pursuit of colonial objectives, suggests Arneil, and in the strategy it adopts, Locke's work is productively understood as part of a tradition of writing defending English rights to American land.

Mathematics and the culture of discipline

These, then, are the lineaments of the seventeenth-century 'age of the improver' as it figures in a broad consensus on the modernity of seventeenth-century English culture. But where does geography fit into this picture? How might the newly mathematized arts of seventeenth-century cartography have expressed and promoted the ethos and ideology of improvement?

In American travelogues, several examples of which we know he owned, Locke found empirical evidence for his argument in the *Two Treatises* that productive property was the just reward of industrious art. Elsewhere he used the same material to make the complementary argument that the proper end of art is the improvement of the human condition, something of which all mankind is capable, but to which only the industrious and rational have attained. There are, he writes in *An Essay Concerning Human Understanding* (completed 1690; first published 1876), 'Nations where uncultivated Nature has been left to it self, without the help of Letters, and Discipline, and the Improvement of Arts, and Sciences' (Locke 1975: 1.4.8, p. 88). Amongst these are the American Indians: 'had the *Virginia* King *Apochancana*, been educated in *England*, he had, perhaps, been ... as good a Mathematician, as any in it' (ibid.: 1.4.12, p. 92).

Locke demonstrates here a distinctly Baconian faith in science, and specifically mathematics, as the mark and means of human improvement, and the dividing line between natural and civil man. Bacon regarded mathematical reduction as the first principle of a rigorous intellectual engagement with and improvement of the material world, recommending 'that all natural bodies be, as far as is possible, reduced to number, weight, measure, and precise definition' (quoted in Webster 1975: 351). It is this general project of a mathematical reduction and re-conquest of the human environment that sociologists of science such as Charles Webster think of as driving a late sixteenth- and early seventeenth-century scientific revolution, and ultimately a revolution in attitudes to property and empire (Webster 1975: 350).

As Webster describes it, the period witnessed an 'increasingly close rapport' between merchants, scholars and artisans which fuelled an array of new mathematical technologies designed for immediate practical use (Webster 1975: 350). In London the new mathematics was promoted by private teachers, public lectures, instrument makers, practitioners, and writers of a burgeoning array of vernacular mathematical texts. Perhaps most famous amongst the latter is John Dee, whose 'Mathematicall Praeface' (1570) to the first English Euclid provided a 'definitive guide to the emergent class of mathematical practitioners, who had attained a firmly established position in the capital by 1600' (Dee 1570; Webster 1975: 349). Further into the seventeenth century the Hartlib circle gave mathematics prominence in their schemes for universal reformation.[3] Mathematics, suggests Webster, gave these architects of reform the confidence and scope to gaze beyond their immediate horizons. Puritan intellectuals regarded accurate measurement as a precondition for 'plantation and profitable exploitation' (Webster 1975: 81).

The technologies of measurement were certainly advanced in their potential and their status by the turn of the seventeenth century. By the second half of the sixteenth century maps were already regarded as 'an essential tool of government' (P.D.A. Harvey 1980: 156). The hegemony of the crown in supplying direct sponsorship of cartography declined in the later sixteenth century from a zenith under Henry VIII as the crown itself declined in financial means, and as other potential sponsors began to foresee the benefits of their own involvement (Barber 1992). Increasingly other, and often better-paying consumers than the Crown were supporting mathematical practitioners not just through patronage and consumption of regional and national topographies, but through employing them privately to survey and map their estates. It is this rise of mathematical surveying and cartography in English estate management which is most often cited as the geographic expression of a new ethos of individualism and improvement. These new practices, we are told, posed a universal mathematical discipline as alternative to the localized ministrations of patriarchal stewardship, confronting and displacing traditions of management and representation deeply entangled with the social landscape of the manorial economy.

Until at least the sixteenth century, the work of surveying involved the 'engrossment', or writing up in detail, of all the legitimate, customary uses of a piece of land, a process of verbal interview and 'auditing' which did not particularly prioritize measurement (Turner 1983: 93). 'Surveie consisteth,' explains Edward Worsop, in a 1582 treatise instructing surveyors in their art:

> upon three principal parts: that is to say the Mathematicall, the Legall, and the Judiciall. Unto the Mathematical part belongeth true measuring, which is Geometrie; true calculation of the thing measured, which is Arithmetike: and true platting, and setting forth of the same to the eye, in proportion and symetrie, which is Perspective. To the Legall part belongeth the knowledge of keeping courts of surveie, of the diversities of tenures, rents, and services, likewise how to make terrors, rentals ... & also how to engrosse books, with many other things appertaining to ye part ... The Judicial part consisteth upon the consideration, and knowledge of the fertilitie, vesture, situation for vent, healthsomeness, commodiousnesse, discommodiousnesse, and such like of every kinde of ground, building, and encrease, in his owne nature, & kind ... Qualitie, and quantitie be unseparable companions.
>
> (Worsop 1582: sig.I3v)

Much of the evidence for a revolution in the practices and the ethos of surveying comes from manuals such as Worsop's: manuals produced by mathematicians to instruct in and promote their knowledge and their profession, and which seem, in one respect, to place a growing trust in mathematics over legal stewardship.

The earliest writers to tackle the subject of surveying focus principally on what Worsop identifies as the 'legal' and the 'judicial', defining the surveyor's work in terms of close empirical contact with his human and material charges, and in terms of verbal inquiry and report (Fitzherbert 1523; Leigh 1577; Folkingham 1610). Edward Worsop warns that only the ignorant think that 'if a Geometer but once looke through the sights in his instrument' he comprehends his subject (Worsop 1582: sig.I4v). However, Worsop is, at the same time, one of the first writers on surveying to stress the importance of a thorough knowledge of geometry. In this insistence on geometrical education Worsop acknowledges and cautiously celebrates the introduction into the profession of new technologies. And in manuals contemporary to Worsop's these new technologies were taking centre stage. Leonard Digges's *Pantometria*, a manual 'finished' and published by his son Thomas in 1571, is exclusively concerned with problems of measurement. Feeling none of Worsop's doubts, Digges is certain that the 'veritie' of his 'experimentes' in mensuration 'shall never be impugned, being so firmly grounded, garded, and defended with Geometricall demonstration, against whose puissance no subtile Sophistrie or craftie coloured arguments can prevaile' (Digges 1571: n.p.).

In later surveying manuals, Digges's preference for geometric foundations begins to manifest itself as a firm commitment. Ralph Agas, in *A Preparative to Plotting of Lands and Tenementes for Surveighs* (1596), argues that survey with geometrical instruments, though 'but new, and scarsely established,' is 'certaine, perfect, and true, without any want or defect' (Agas 1596: 2). Agas is explicit in comparing geometric survey with verbal means of textualizing land. The instrumental method 'to the said use of surveigh,' he asserts, is 'of all other devices by bookes or otherwise the most sure and lasting' (Agas 1596: 2). Whereas many abuses may be suffered in a manor which 'their auncient and faire bookes' are unable to prevent, he insists, 'the surveigh, by plat, suffereth no such inconvenience' (ibid.: 15). Aaron Rathborne's *The Surveyor in Foure Bookes* (1616) makes geometrical theory; geometrical techniques; and geometrical instruments the principal subject of his treatise (Richeson 1966: 202–3). In his privileging of universal geometry over quality and law, Rathborne presents the pattern for most survey books of the rest of the seventeenth century (see Leybourn 1653; Love 1688). In these surveying manuals the manor no longer features as the almost exclusive stage on which surveyors act, displaced by a generalized Cartesian continuum.

This reduction of the intricacies of manorial stewardship to disciplined mathematical form has been interpreted widely as the symptom of and precondition for a new theory of economic and social value rooted in individual endeavour and improvement. For Bernhard Klein the 'upstart art' of land measurement reflects and forwards a shift from a conception of agrarian space as a 'social realm' to an understanding of it as 'marketable commodity' (Klein 2001: 44). Moreover, as an integral element in the *translatio imperii*, historians such as Klein find the estate surveyor's cartographic commodification of space transformed into a powerful practical and ideological tool for appropriating land abroad. Bruce McLeod reads estate surveying alongside house rebuilding, town planning, sixteenth- and seventeenth-century utopian writing, country house poetry, literary epic, American travelogue and captivity narrative as products of the same expansionist imaginary, structured by the same explicitly or implicitly geometrical view of material space. In a chapter on Spenser, McLeod writes that the English:

> treated Ireland as though it were *terra nullius* and thus easily and geo-
> metrically subdivided into territorial units. The conception of space as
> *absolute* – as an empty field or container – derives from Euclid's geo-
> metry ... introduced to England for the stated purpose of enhancing
> imperial power ... Absolute and abstract, the geography of Ireland
> became so much property value, a quantity more readily dealt with and
> dealt out by the English.
>
> (McLeod 1999: 53)

In accounts such as McLeod's, the practices and aesthetics of mathematical reduction and discipline in surveying, cartography and Baconian science in general harmonize closely with the distinctive values and aesthetics of Puri-tanism and so quickly accumulate legitimacy in the radical climate of mid seventeenth-century England. Much sociological scholarship on seventeenth-century Puritanism has suggested that it is better understood as a distinctive and unifying set of cultural practices than a distinctive and unifying theology: practices typically (though not exclusively) 'disciplinary' or 'reduc-tive' in nature, and thereby hospitable simultaneously to economic reform and Baconian science (see David D. Hall 1989; Peterson 1997).[4] In the early seventeenth-century Puritan campaign for the 'reformation of manners' which provoked James I's 1618 *Declaration of Sports*, historians from Max Weber to Christopher Hill have found their most persuasive evidence for a characterization of Puritanism as the ethos of the individualist middling sort, rationalizing 'their efforts to discipline the poor, to curb their drunken, promiscuous ways, and to instill in them respect for sobriety, property and hard work' (Spurr 1998: 76).

When James I wrote to protect rural sports he was defending a moral vision that included the whole customary order of the post-feudal economy, bound together by principles of stewardship, hospitality and beneficence as well as duty and obligation. Against these traditional, customary values Puritan discourse advanced a kind of moralized aesthetic: one of reformation, discipline, reduction. In this disciplinary Puritan aesthetic Christopher Hill finds the basis of a 'labor discipline ... felt by the non-working classes to be a national necessity, preached now by economists with the same zest as by theologians' (Christopher Hill 1980: 253). 'Through penitent labour,' sum-marizes Charles Webster, 'degenerate man was allowed to evolve an ordered civilization in which the human condition was ameliorated to a certain degree', and if the value of such labour depended on such discipline it justi-fied its rigorous imposition (Webster 1975: 325). Bruce McLeod, in turn, finds this will to discipline expressed in domestic as much as explicitly colo-nial writing; in country house poetry and mathematical surveying as much as Puritan colonial history (McLeod 1999). In the obsession with spatial discipline which he finds in early Puritan writing, in particular, McLeod identifies the perspective of 'yeomen, artisans, and gentry', concerned to organize material and social space in the interests of improvement, productivity, and their own benefit (ibid.: 108–14).

McLeod is typical of recent cultural historians in regarding a Puritan culture of discipline as the ground-plot not just for early modern capitalism, but also for the colonial projects of the seventeenth century. As canonical studies of Puritan intellectual history have made abundantly clear, the American Puritan imagination was characterized by a distinctive spatiality, importing with it such concepts as the 'city on a hill', and the community 'paled' in against the corrupt wilderness of the world. Puritan histories of the early settlements define their communities in terms of fences, hedges, pales, and express their nagging paranoias about political and theological disunity in equivalent spatial figures of encroachment and scattering. In McLeod's account, as in many others, a Puritan concern with order and containment, often expressed in explicitly mathematical and indeed cartographic terms, merges fairly seamlessly with a Lockean doctrine of improvement and a mercantilist impulse to exploit and commodify colonial land.

With this sweeping assessment of the ideological uses of early modern geometry and geography we have come full circle, returning to that thesis expressed in exemplary form by Brian Harley. For Harley seventeenth-century cartography 'fostered the image of a dehumanized geometrical space ... whose places could be controlled by coordinates of latitude and longitude'; served as a means whereby colonial land 'could be appropriated, bounded, and subdivided'; and went hand in hand with a Lockean doctrine of rightful appropriation through improvement (Harley 2001: 189, 190). 'Maps with their empty spaces,' writes Harley of the seventeenth-century mapping of New England, 'can be, and were, read as graphic articulations of the widely held doctrine that colonial expansion was justified when it occurred in "empty" or "unoccupied" land' (ibid.: 190). Yet once again this seamless merging of capitalism, imperialism and Puritan discipline obscures some crucial and defining faultlines in seventeenth-century English culture.

Limiting the age of improvement

I think we often greatly exaggerate the confidence and consistency of a seventeenth-century culture of mathematical discipline and improvement. At its worst, the case for the seventeenth century as the 'age of the improver' rests on little more than intuitive analyses of the empty spaces and panoptic geometries of early modern maps. At its best it still tends to be supported by a carefully plotted route through the most 'modern' aspects of seventeenth-century social and scientific reformism and professional propaganda. Like traditional histories of early modern cartography this account turns a lengthy, doubtful and contingent journey into a metamorphosis, eliding processes of conflict and negotiation. From this point on I want to plot a different and more complex route: a route which navigates the unstable currency of early modern mathematics; the uncertain and fragmented politics of early modern geography; and the inherent ambivalence of Protestant humanist culture towards artful individualism and economic change.

The tradition of scholarship still thrives which regards Puritanism primarily as an ethos of the middling and better sort, deployed as an instrument of social control and functioning as the midwife of entrepreneurial capitalism (see, for instance, Innes 1995). Yet the most recent research has also confirmed the 'obstinately religious' nature of Puritanism; the vertical penetration of Puritan commitment throughout the class spectrum, not least through the medium of cheap print; and the preparedness of even the best placed early seventeenth-century Puritans not just to work through established institutional channels to secure reform, but to challenge the ecclesiastical and civil establishment to their own material cost (Durston and Eales 1996: 11; Spufford 1995; Tessa Watt 1991; Spurr 1998: 78). Above all, we must be careful not to underestimate the moral anxiety fostered by Calvinist theology, with its revulsion towards human nature and the world, redeemed only by the hope of saving grace. As a form of 'anxious Protestantism', differing more in intensity than in doctrine from its root form, Puritanism rendered this anxiety peculiarly intense (Durston and Eales 1996: 11–12). 'For the Puritan,' writes Ernst Cassirer, 'that peculiar oscillation prevails in respect to the world, that alternation between asceticism and attraction' (Cassirer 1970: 69).

Even the most pragmatic, worldly streams of Puritan and Baconian thought, equating truth and virtue with utility, contended over a long period with a residual discourse of complaint, associating practical art and economic individualism with moral and social corruption. Treatises promoting improvement and economic reform were matched throughout the mid-seventeenth century by sermons and print diatribes denouncing enclosure and depopulation (see, for instance, Powell 1636; John Moore 1653). This was a period within which the legitimacy of all customary land-use, including private property, vied with the legitimacy of the absolute right of property, not just across English culture, and between conservative court/peasant and progressive, Whiggish interests, but even within the ideas and discourses of individuals representing and debating the use of land. This was a period not just of contestation between polemicists expressing opposing economic views, but also of a more general instability and indecision between traditional and new ideas and values.

It is possible to interpret the Puritan culture of 'discipline' or 'reduction' in two, apparently contrary ways. In its war on custom the Puritan ethos may well have played a key role, as the 'Weber thesis' suggests, in dismantling the Tudor moral economy whose conservative values persisted into the seventeenth century (Weber 1930). On the other hand, Puritan reductionism can be taken largely at its word in its many tirades against worldly 'declension' into vanity and covetousness, as an expression of a spirituality inherently at odds with the gathering tide of modernity.[5] But a third, and I think most convincing, mode of interpretation finds Puritan cultural practices negotiating the characteristic ambivalence of the Puritan mentalité between *Contemptus Mundi* and the world. Practices such as fasting, and a cultural aesthetic of discipline, acted as a 'counterweight', thinks Christopher Grasso, within societies coming to embrace the Puritan war on custom, 'to attitudes and structures encouraging acquisitiveness and economic individualism' (Grasso 1999: 202). It is in the light of this careful rhetorical balancing and negotiation that the following three chapters will suggest a re-consideration, in turn, of the cultural currency of seventeenth-century mathematics; of the mathematical aesthetics and rhetorics of a Protestant culture of discipline; and ultimately of the mathematized discourse and aesthetic of seventeenth-century geography.

3 Humanist geometries

Circles in the sand

Limiting Baconian science

My account of a seventeenth-century mathematical discourse and aesthetic entangled in the ambivalence of a Protestant culture of discipline runs counter to the strong-headed pragmatism predominantly emphasized in the recent cultural history of early modern science. 'In recent years', writes Lesley Cormack, in an article on early modern English geography, 'we have begun to recognize the centrality of practice in defining science' (Cormack 1991: 639). Practice is very broadly conceived here. The work Cormack credits with this growing insight ranges from Robert Merton's, Charles Webster's and Christopher Hill's scholarship on Puritanism and science, through E.G.R Taylor's and J.A. Bennett's histories of mathematical technology, to Steven Shapin's explorations of seventeenth-century experimental culture. What these diverse histories have in common is their situation of theoretical science in relation to something beyond pure intellectual theory. In Taylor's and Bennett's work, theory addresses immediate practical problems and both depends upon and is conditioned by the evolution of instrumental technology. In Merton's the broad social context of English Puritanism conditions, if not 'the form or content of scientific knowledge or scientific method', then at least 'the dynamics and social standing' of science, and specifically the seventeenth-century 'upsurge of interest in and approval of science and technology' (Shapin 1988b: 393). Because of Merton's ground-breaking thesis, writes Steven Shapin, 'no historian now seriously maintains that the thematics and dynamics of scientific activity ... are unaffected by social and economic considerations' (ibid.: 604).

The histories Lesley Cormack cites, from Merton to Shapin, militate against a conception of early modern science as disinterested, and indeed reveal the 'artificiality' of any historiographic separation of early modern science into theory and practice (Cormack 1991: 639). In Cormack's article and in her subsequent monograph geography is 'emblematic' of the true nature of the 'new science'; its development conditioned by the broadly conceived social goals of national self-definition and imperial expansion; and its values profoundly civic and pragmatic (Cormack 1997; Cormack 1991:

640–1). Cormack writes: 'the study of surveying in England was fundamentally practical ... eschewing any theoretical discussion ... aimed at an audience that was practical rather than abstract and so was guided by the marketplace rather than by the commonwealth of ideas' (Cormack 1997: 168). Yet just as this characterization totalizes the politics of what was in reality a necessarily ambivalent and heterogeneous set of geographic practices, I think it also reifies the epistemologies of seventeenth-century science. Baconian pragmatism and empiricism was only one strand in the unstable cultural currency of geography and the mathematics that informed it. It is my contention that this very instability fitted mathematics as a rhetoric simultaneously of material profit and ideal value; atomist individualism and civil limitation; and that made mathematics the core language and aesthetic of the Protestant culture of discipline.

The trenchant assertion in Cormack's work of the utilitarian, Baconian nature of early modern geography is designed in part to revise an earlier trend in the twentieth-century historiography of early modern science which I think deserves rather more accommodation. The Merton thesis does little to trouble the established teleology of scientific progress. Older intellectual histories look to the seventeenth century – the age of Bacon, Boyle and Newton – for the birth of a modern science which displaced the laws of traditional authorities with a new attention to the particular entities and attributes of nature. The challenge posed by the 'Merton thesis' to this teleological history of science was to suggest that the pragmatic and empiricist developments it presented as necessary; points on a trajectory from error to truth; from sterile scholarship to public service; may have depended on the contingent circumstances and events of a particular society. But another kind of challenge was also posed in the second half of the twentieth century which questioned this trajectory itself.

From the 1960s onward Frances Yates and her associates at the Warburg Institute sought to revise the history of Renaissance science as the birth of modern empiricism and civic humanism. Yates stressed instead the contemporary significance of what now look like 'dead ends' in the history of science. She emphasized an alternative, Platonic strand in early modern scientific culture which was anti-empiricist and esoteric. The Yates thesis opened up a field of research which traced the influences of neo-Platonism not just in figures such as Cornelius Agrippa and Giordano Bruno, whom a teleological history of science considers relatively marginal to the scientific revolution, but in some, such as Boyle and Newton, traditionally considered central. As Yates puts it in *The Art of Memory* (1966), 'the Renaissance conception of an animistic universe, operated by magic, prepared the way for the conception of a mechanical universe, operated by mathematics' (Yates 1966: 224).

The Yates thesis makes room for the idealism of intellectual traditions which owed substantial debts to Plato. The humanist cultures of Italy and Northern Europe may have been civic, but they also promulgated both an explicit neo-Platonism and the idealism implicit in Aristotle. In J.G.A. Pocock's words, humanism embraced the paradox that whilst 'only the life of the citizen was truly ethical and human', 'only the abstract world of unmotivated contemplation was truly intelligble and real' (J.G.A. Pocock 1975: 56). 'The humanist,' writes Pocock, 'was ambivalent between action and contemplation; it was his *metier* as an intellectual to be so' (ibid.: 59). Whilst English humanism characteristically argued that all knowledge should refer itself to human use, it followed Aristotle in privileging abstract over practical science as a better form of knowledge. Accordingly the classical quadrivium, which formed the basis for sixteenth- and seventeenth-century school and elementary university teaching in four of the seven liberal arts, divided geometry, arithmetic, astronomy and music into the hierarchy of speculative and practical, implying the superiority of knowledge divorced from practical use. At university mathematics was principally a preparation for advanced study in the three philosophies: Natural, Metaphysical and Moral (Jesseph 1999: 74).

The tensions between a residually Yatesian history, which foregrounds the ambivalence Pocock describes, and a new sociology of science, which tends to overlook it in its emphasis on Baconian pragmatism and empiricism, have manifested themselves in recent years in revisionist accounts of key contributors to early modern scientific culture. Significant amongst these has been the mathematician John Dee. Yatesian history encourages us to recognize two aspects to sixteenth- and seventeenth-century mathematics: one idealist, seeking esoteric means for cultivating magical intellectual powers; the other empiricist and pragmatic. It has identified these two currents with two giants of Renaissance educational reform: Dee, the English magus and propagandist of British 'empire', and the Huguenot martyr Pierre de la Ramée, known often to his contemporaries in the European scientific community by his half-Anglicized, half-Latinized name, Peter Ramus. Both of these figures were responsible for publications which sought to broaden the audience of mathematics beyond the universities where it was felt to have fallen into neglect: Dee in his 'Mathematicall Praeface' to the first English vernacular edition of Euclid's *Elements*, translated by Sir Henry Billingsley and published in 1570; Ramus in his *Geometria* (1569), and in his published lectures and addresses on mathematics.[1] Moreover, the influence of these figures was not channelled only through their own much-reprinted publications. Dee's 'Praeface' attended other editions of Euclid over almost a century following its original appearance (French 1972: 172–7), and Ramus's *Geometria*, although excluded by other popular texts discussed below from universal usage in England, was cited and sampled with great frequency for its doctrines and its organization by other mathematical publicists (Hooykaas 1958: 106–18).

A brief reading of the texts these mathematicians published encourages the Yatesian idea of contrary intellectual currents. Where he deals with the origins and nature of mathematics Ramus discovers a humanist 'closed circuit' leading from and back to human use (Hooykaas 1958: 20–1).[2] Ramus divides science into the successive stages of nature, art (which follows nature) and practice (which follows art) (ibid.: 22). In accordance with this sequence, the principles of any art are discovered not a priori, but through study of its usages and through rational reduction of these to generalities (ibid.: 23, 51). Hence Ramus famously embraces a definition of the science as *'ars bene metiendi'* ('the art of measuring well') (Ramus 1569: 1). Ramus bases his trust both in the rightness of usage and in the possibility of knowing usage rightly on his doctrine of 'natural reason', by which he means the inherent capacity of even, and perhaps especially, uncultivated human intelligence to follow nature, forming correspondences with the natural order of the universe (Hooykaas 1958: 51). This ability to follow is strictly not an ability to precede: Ramus acknowledges no Platonic 'innate' ideas. In turn, John Dee's 'Praeface' promotes a whole range of practical mathematical arts, from geography to architecture: what Dee calls 'Artes Mathematicall Deriuatiue' (Dee 1570: sig.a.iiir). It does so avowedly to improve the work of craftsmen and to benefit the wider commonwealth. This is, in other words, a civic, pragmatic text. But where Ramus discovers the origins and essential nature of mathematics in human use and embraces the name 'geometry', Dee describes what he calls 'mathematicals' as so 'free from all matter' that another name than 'Geometry' 'must needs be had' for a science which 'regardeth neither clod, nor turff: neither hill, nor dale' (Dee 1570: sig.*1r, sig.a.2v).

What should we make of this difference between two such fathers of early modern science? Frances Yates herself characterizes Dee as an embodiment of the paradoxes of the English Renaissance: a thinker inspired by the contemplative philosophy and esoteric objectives of Hermetic neo-Platonism, yet at the same time conditioned by the civic tone of Latin humanism to deploy his knowledge in the service both of the court and of a national community of artisans (Yates 1969: 1–19). We might consider this paradox further embodied in the social relationship of two apparently contrary thinkers. Despite their variant philosophies of science, Dee and Ramus appear to have perceived themselves as part of a unitary enterprise of reform (French 1972: 167). They corresponded as 'friends', apparently concerning mathematical texts, and Ramus even sought to intervene with the Crown on Dee's behalf to gain him a university chair in mathematics, which both Oxford and Cambridge at that time lacked (French 1972: 142; Hooykaas 1958: 105). The two men were in fact part of an emerging network of social relationships and knowledge transactions through which the cause of mathematical science in the late English Renaissance began to be advanced: a network energized not just by broad religious sentiments or economic dynamics but by immediate practical, public goals. Dee was not just a propagandist for an English empire founded on navigation and other reformed mathematical arts; he advised the pioneers of English colonization directly and held a stake in an early patent (Wallis 1985: 29). And it is precisely in the light of this localized social context that revisionist accounts of Dee have disputed the Yatesian reading of his writing and career, discounting his idealist departures from civic humanism.

Within Lesley Cormack's own taxonomy of early modern geographic practice, even John Dee's apparently neo-Platonist account of mathematics and its derivate arts is placed within the context of an over-arching utilitarian ideology. In William Sherman's 1995 study of Dee this broadly pragmatic assessment of Dee is refined to the same fine detail characteristic of Steven's Shapin's cultural geography of experimental science. Sherman's book positions itself explicitly as a revision of the Yatesian 'interpretative myth' of polar intellectual cultures (Sherman 1995: 12). Sherman argues that the division between neo-Platonism and Hermeticism on the one side, and civic humanism and Baconian empiricism on the other, may make sense to modern intellectual history, but is not in any clear sense an early modern division (ibid.: 20). In place of this intellectual history, Sherman offers a context for John Dee's work that takes into account Dee's self-fashioning to fit distinct social roles: characteristically the marginal roles of mediation between powerful and wealthy individuals and groups (ibid.: xiii). The self Dee fashioned for these roles, Sherman suggests, incorporated elements both of neo-Platonism and humanism (ibid.: 14–15, 21, 25). If this combination seems paradoxical, this is a modern paradox, and not an early modern one (ibid.: 22–3).

I very much concur with Sherman's assessment of the scientist in early modern England as, to quote one of his early modern sources, 'a Factor and Trader'; a mediator and gate-keeper of knowledge (quoted in Sherman 1995: 39–40). Sherman is right to resist seduction by the Yatesian myth of the solitary, eccentric philosopher: one which has dominated both academic and popular perceptions of Dee and his ilk. Yet as Sherman himself acknowledges, this myth is not unique to historiography, but was perpetuated by early modern self-fashioners themselves. It is a version of that rhetoric of 'solitude' deployed in cultures from classical Greece to European Romanticism to support claims to intellectual authenticity. Steven Shapin finds Robert Boyle, for instance, presenting himself and celebrated by his contemporaries as 'an intensely private man', and refers this rhetoric to a traditional association of 'valued knowledge' with spatial isolation: the isolation of 'the monastic cell and the hermit's hut' (Shapin 1988a: 384). Boyle struggles, in Shapin's account, to reconcile this spatial model of epistemic value with a simultaneous humanist impulse to attack sequestered scholarship. Shapin's account of Boyle suggests that if Dee fashioned himself as a solitary philosopher, he did so not because that was what he was, but because this traditional image enhanced the value of his intellectual stock in a period where intellectual values were in flux.

Whilst I agree with Sherman's argument about self-fashioning, I think it replaces Yates's artificial separation between esoteric neo-Platonists and civic humanist empiricists with its own artificial separation between what people said and what they did: the inside and the outside of the written text. This artificial separation between texts and lives is the same one Barbara Belyea brings to light in Brian Harley's work. It has been underscored by Lesley Cormack, who is resolute in trying to resolve the contradiction between John Dee's textual idealism and his social pragmatism, to the disadvantage of the idealist text.[3]

I think the 'Yatesian' tension Sherman and Cormack perceive between the pragmatic lives and idealist texts of Dee and early modern geography itself should not be dispensed with so quickly. I think that it's a tension that remained central to early modern scientific culture for most of the seventeenth century, as Douglas Jesseph's recent study of a seventeenth-century mathematical controversy confirms (Jesseph 1999). On one side of the battle reconstructed in Jesseph's book stands Thomas Hobbes; on the other, the Oxford University triumvirate of Seth Ward, Savilian professor of astronomy; John Wallis, Savilian professor of geometry; and John Wilkins, warden of Wadham. Print skirmishes between Hobbes and the Oxford triumvirate were fought over everything from Hobbes's denial of the immateriality of the soul to his flawed attempts to solve classical mathematical problems such as the quadrature of the circle. The Oxford mathematicians expressed vehement opposition to the foundation of Hobbes's philosophy of science, upon which he built his subsequent philosophies of humanity and society: a materialist mathematics.[4] In the course of a long, bitter and very public dispute John Wallis vigorously attacked Hobbes's mathematical materialism and defended the integrity of pure mathematics, remaining, in Jesseph's words 'firmly committed to the view that abstract mathematics deals with objects not found in nature' (Jesseph 1999: 133).[5] In both social and intellectual terms the Oxford triumvirate prevailed in this dispute, leaving Hobbes humiliated and discredited as a mathematician. In the closing chapters of his book Jesseph draws conclusions which he feels conflict with the 'sociological reductionism' he associates with the Shapin school. Wallis's idealist and Hobbes's materialist positions seem to reverse the correspondences Schaffer and Shapin make with their respectively moderate and absolutist politics, undermining their explanation of Hobbes's exclusion from the seventeenth-century experimental community (ibid.: 352). Jesseph aligns himself with another, more 'nuanced' explanation (ibid.: 275). As a materialist bogeyman, Hobbes offered his experimental peers an opportunity to distance themselves from the materialist implications of their own work, and to negotiate a moderate scientific role in the public medium of print.

It is my argument in this book that the tensions manifest in the battle between Hobbes and Wallis, and the public print arena within which they were played out, were core to the early modern culture of mathematics. Rather than seeking to define a stable mathematical ethos in the seventeenth century, whether empiricist, civic and pragmatic, or esoteric, liberal and idealist, I think we should be prepared to understand mathematics as a flexible mode of cultural communication. A mode whose unstable meaning rendered it the perfect medium for negotiating the ambivalence of Protestant humanism in the capitalizing culture of seventeenth-century England. In these negotiations written texts were not a sideshow, but the heart of the action.

The double currency of early modern geometry

My intention in this book is to demonstrate two things about the relationship between geometry and what is often rather anachronistically distinguished as 'literary' writing:[6] that the ideas developed in specialist mathematical writing had a cultural currency beyond such scientific communities as can be defined in the late sixteenth and seventeenth centuries, and that the rhetorical trapes through which literary writing dealt with geometry played a constitutive role in negotiating its nature and value, not just in poems, plays and literary prose, but in the most specialist mathematical publications themselves. More concisely, I want to demonstrate the way in which the social currency of early modern geometry and the social credit of the mathematician were constituted rhetorically and circulated through textual discourse. Where we underestimate the role of writing we are likely to conclude with Lesley Cormack that mathematical interest and knowledge in the late sixteenth and early seventeenth centuries was of a 'closed and almost recondite nature' (Cormack 1997: 124). But where we consider the currency of mathematics to be more than a matter of knowledge narrowly and scientifically defined, then the currency of mathematics is considerably broader. This wider discursive currency is, I think, the most important consideration in judging the social purchase of mathematical art. For mathematics to 'work' in early modern culture, for it to have meaning and value for those whose lives it touched, what counted was not who understood the principles of trigonometry, but who used mathematics as a mode of cultural communication.

The first thing to observe about the currency of geometry in early modern literature is that two kinds of geometry were in circulation. Not just two versions of geometry as a discipline, although early modern education divided mathematics into speculative and practical, as we still do into pure and applied. But two essentially different ideas of what mathematics is; where it comes from; and what it's for. On the one hand mathematics was understood as something ideal, Platonic, removed from everyday life. On the other it was understood as inherently worldly and pragmatic. Literary discourses, in other words, followed the contours of those distinct philosophies of mathematical science revealed in the writings and controversies of Dee and Ramus; Wallis and Hobbes. In early modern literary writing these two different conceptions of what mathematics is remain constant whether the ideal or worldly nature of mathematics is being celebrated or attacked, attitudes often conditioned by genre. Geometry functions in literary discourse as a flexible figure; a mode of cultural communication used to attack or embrace idealism or the world, and most importantly, in its very double nature, to negotiate the inherent ambivalence of Protestant humanism.

Some early modern writers dealing poetically with the spiritual aspect of human life characterize geometry and its arts as touchstones of constancy and truth. This is just a passing metaphor in an elegy by Francis Quarles (1637), which pits 'the just Geometry of my Pen' against the calumnies of worldly interest and corruption ('An Elegie upon my Deare Brother, The Jonathan of my Heart', Quarles 1967, III: 15). It is elaborated into emblems several times in George Wither's *A Collection of Emblemes* (1635) (Figure 3.1). And it is a philosophical thesis in the third book of Henry More's 'A Platonick Song of the Soul' (1647), which seeks to prove the immortality of the soul through its perception of innate geometric ideas independent of worldly experience and sense ('A Platonick Song of the Soul', More 1647: 238–9).

But whilst it is possible to find instances of seventeenth-century writers embracing a geometry conceived of as ideal, such a conception is more visible in light-hearted satires. If devotional and philosophical poetry were natural homes for an enthusiastic celebration of geometric Platonism, on the conventionally more worldly stage we find such idealism routinely mocked. In John Fletcher's *The Elder Brother* (composed *c.*1625; first published 1678), the bookish son of the title is threatened by his father with disinheritance for his Dee-like refusal to meddle with 'the dirt and chaff of Nature' (*The Elder Brother*, Beaumont and Fletcher 1905–12, II: II. i, p. 12). The father, Brisac, defends his suspicion of such impracticality, asking:

> Can History
> Cut my Hay, or get my Corn in? And can Geometry
> Vend it in the Market? Shall I have my sheep kept with a
> *Jacobs-staff* now?
> (Beaumont and Fletcher 1905–12, II: II. i, p. 14)

Eustace, younger brother and courtly foil to this scholar, uses geometry in similar figurative vein, this time to express his own, more sophisticated materialism. His mistress's dress, jokes Eustace, is 'not made/by Geometry' (ibid.: II. ii, p. 17). This last ironic joke, rooted in the notion of the incorporeality of geometry and the unworldliness of geometric study, is commonplace on the seventeenth-century stage. In plays by Shackerley Marmion, Robert Davenport and Margaret Cavendish we find cowards who fight by geometry (i.e. not at all), and a beggar's clothes and a worldly globe which, having no support, 'hang by geometry' (*Holland's Leaguer*, Marmion 1875: V. iv, p. 91; Davenport 1661: IV, pp. 36–7; *The Sociable Companions*, Cavendish 1668: front matter, n.p.). In *The Commonwealth of Oceana* (1656) James Harrington identifies this jest on geometric Platonism not just as common but as a cliché amongst the uneducated. Mounting one of his many attacks on Hobbes, Harrington suggests the impossibility of establishing monarchies on anything but ancient custom, 'unless like Leviathan you can hang it (as the Country fellow speaks) by geometry' (Harrington 1992: 56).

By Knowledge *onely*, Life *wee gaine*,
All other things to Death *pertaine*.

ILLVSTRATIO I.   *Book.* I.

Figure 3.1 George Wither, '*By* Knowledge *onely*, Life *wee gaine*, All *other things to* Death *pertaine*', in *A Collection of Emblemes, Ancient and Modern* (London, 1935), 1. Reproduction courtesy of the British Library (shelfmark: c.70.h.5).

Jests about the idealist nature of geometry can seem double edged. They are made at the expense of such useless study, and they are presumably made on the stage, in this respect, to make ignorant people laugh. But another laugh is being had at the expense of those who, like Brisac and Strepsiades, are too stupid to appreciate the value of liberal science. Harrington's is a complex insult, apparently associating Hobbes's materialism with the bumpkin's muddy ignorance. A materialist geometry, he seems to imply, is indeed a nothing from which to hang a something.[7] But however ambiguous the value of geometry in these various instances, there is an agreement about its liberal, ideal nature. And yet concurrent with this idealist understanding of geometry runs another thread in early modern literary discourse, which understands geometry as wholeheartedly pragmatic. As with idealist geometry, pragmatic geometry is as visible in satires against it as in celebrations of it.

If a gentle satire on geometric Platonism was an incidental cliché both on the early modern stage and, if we are to believe Harrington, in common, 'country' discourse, then a still greater cliché, which typifies the treatment of geometry and geography in early modern poetry, involves a more serious rejection of the opposite: geometric worldliness. Robert Appelbaum and Richard Helgerson have both written of a persistent vein of rhetoric which runs counter to interest in and valorization of the new Renaissance geography, debunking it as folly and producing extravagant fictional geographies which undercut the claims of science (Appelbaum 1998; Helgerson 2001). This hostility to geography, a product of the same *Contemptus Mundi* expressed in Protestant complaints against enclosure, extends to the mathematics by which early modern geography was increasingly informed. Meditating on scripture, and making assumptions about geometry the polar opposite of his contemporary Henry More's, R. Fletcher mocks the inadequacy of a worldly science to define the constant spiritual:

> True blessed *Saviour*, true! thy Kingdom's not
> Of this world. For we cannot finde a spot
> Of thy *Crown Land*, where *Geometrie* may stay
> Her reeling compass to move any way
> In demonstration of that circling Round
> That may define th'inclosure *Holy ground*.
>   ('John *chap*. 18. *ver*. 36. *My Kingdom is not of*
> *this World*', Martial and Fletcher 1656: 221–2)

In George Sandys's translation of Ovid's *Metamorphoses*, Fletcher's association of geometry with the 'inclosure' of material property is confirmed as inherent and original, rather than merely a misapplication of a blameless art. Here the advent of geometry is identified with the 'brazen age' of human history, and the arrival of war, fraud, treason and property: 'The Ground, as common earst as Light, or Aire, / By limit-giuing Geometry they share' (Ovid 1970: I. 135–6, p. 29).

Any criticism paying attention to the determining influence of genre would quite properly warn us about the formulaic nature of such poetic satires on human vanity, and by extension about their dubious relevance to the everyday currency of early modern geometry. Fletcher and Sandys rehearse and translate conventional devotional and Pastoral rejections of worldly art. But whilst these poetic assessments of geometric value may be formulaic they appear to draw upon a fund of assumptions about the nature of geometry which exceeds such formal positions. We might doubt their sincerity in rejecting the uses of geometry, including the definition of private property, but we ought to take seriously their presumption of its pragmatic nature.

Just as the ideal nature of geometry is both satirized and celebrated and – between these two positions – taken for granted, the same is true for a geometry understood as definitively of the world. Where Sandys's Ovid laments the historical advent of geometric surveying as symptomatic of the fall from a golden age of commonality, an equally common narrative celebrates its historical birth in Egypt, traditionally the nursery of civilized human arts.[8] Herodotus, Englished in a 1584 edition attributed to Barnabe Rich, is a key Renaissance source of this myth of origins:

> The same King [Sosostris] made an equall distribution of the whole countrey to all his subjects, allotting to every man the lyke portion and quantitie of ground, drawne out and limited by a foursquare fourme. Heereof the King himselfe helde yeerely revenewes, every one being rated at a certayne rent and pension, which annually he payd to the crowne, and if at the rising of the floud it fortuned any mans portion to be overgone by the waters, the King was thereof advertised, who forthwith sent certain to survey the ground, and to measure the harmes which the floud had done him, and to leavy out the crowne rent according to the residue of the land that remained. Heereof sprang the noble science of Geometry, and from thence was translated into Greece.
>
> (Herodotus 1924: 196)

We find reference to this myth of geometric origins in many places in Renaissance writing. Even Dee tells the story, conceding that notwithstanding the essential detachment of geometry from measurement 'the first and notablest benefite' shown by geometry to man may well have been the re-establishment of Egyptian 'Boundes and meres', and, 'when disorder preuailed', the distribution of 'Commons' into 'seueralties' (Dee 1570: aiir). One of the most interesting instances of the myth is in a 1652 translation by Francis Goldsmith of Hugo Grotius's tragedy *Sophompaneas*, a tragedy telling the biblical tale of Joseph's service to Pharaoh. In the third act of Grotius's tragedy Joseph's brothers Simeon and Judah are reflecting admiringly on Joseph's governorship of Egypt. Joseph has proved himself a patron of the arts, liberating some of Pharaoh's subjects from taxation in order that they may study religion, cosmology and geometry. 'See you those figures written in the dust?' asks Simeon:

> How great a surface a triangle must
> Contain, they dispute: what proportion there
> The Circle hath to the Diameter.
>
> (Grotius 1652: 27)

Goldsmith, who states that the information for most of his notes comes from Grotius himself, or his dedicatee Gerard Jan Vossius, annotates these lines with an account of the Egyptian invention of geometry:

The old Ægyptians much exercised themselves in Geometry, Arith-
meticke, Astronomy; Geometry they studied of necessity, because when
the bounds of their fields were removed by the o'reflowing of *Nilus*, after
the River was returned into it's channel, every one was to have their
owne restor'd again: nor could the portion of grounds be otherwise
assigned then by applying an art, which made up out of certaine principles, could not deceive skilfull measurers. And it was therefore called
Geometry, as it were a measuring of the Earth.

(Grotius 1652: 81–2)

Hugo Grotius, as we have seen, advanced a theory of natural law as a basis
for rational human government: a theory highly influential on Locke. Here
Grotius figures geometry as a bridge between natural law and human
society: derived from nature, yet eminently suited to the consensual government of civil human life. This myth is a crucially important one both for a
humanist philosophy of mathematics and for the evolving legal and economic philosophies of the seventeenth century. Where Ovid identifies un-
geometric commonality as a 'nature' from which geometric surveying
departs, in Grotius's rehearsal of the Egypt narrative private property and its
boundaries are a basic form of human order which geometry is called upon
to restore: a form of use so elementary that it can ground both science and
law.

Humanist geometries: circles in the sliding sand

This Egyptian narrative about the humble practical origins of geometry was
not always told to demonstrate the essential and necessary pragmatism of
geometric knowledge once refined to abstract principles. But many early
modern writers do celebrate precisely the humanist, pragmatic spirit of a
geometry both derived from and intended for human use. A poem in Henry
Peacham's emblem book *Minerva Britannia* (1612), for instance, advises:

> Geometry or wishest thou to learne,
> Obserue the Mill, the Crane, or Country Cart,
> Wherein with pleasure, soone thou shalt discerne
> The groundes, and vse of this admired Art,
> The rules of *NVMBRING*, for the greatest part,
>   As they were first devis'd by Country Swaines,
>   So still the Art with them entire remaines.
> ('Rura mihi et silentium', Peacham 1612: 186)

The poet and topographer John Norden, a figure whose corpus appears so
heterogeneous to modern eyes that historians have surmised two Nordens,
also champions geometry in a devotional poem published in 1600 as an art
developed empirically, by the 'proofe' of material practice:

Geometrie the mother of all Arts,
Was not at first found by a former Art:
Nature did first deliniate those parts,
That *Wits* and *Willes* might come vnto her mart,
   And buy by practise (to adorne the heart)
      The principles of Art, as *Archimedes* did,
   Archytas too, and other, to some hid.
            (Norden 1931: st. 144, n.p.)

So far I've established the currency of opposing views of the nature of geometry, consistent between positive and negative evaluations of such natures which seem partly determined by genre. Devotional poetry, for instance, generally consistent in its rejection of a changeable world, variously embraces or scorns geometry as escape-route or worldly trap. But not all figurations are so self-consistent. John Norden clearly commits himself to a geometry original in practice: a geometry of the 'mart'. However, in his insistence upon geometry as the 'mother of all Arts' and indeed adornment of the 'heart' he might be suspected of trying to have it both ways.

Norden's uncertainty between ideal geometry and a geometry of the world finds more extensive expression in a late sixteenth-century survey of human arts. The Huguenot soldier Guillaume de Salluste du Bartas's epic chronicle of the creation *La Sepmaine* was published first in 1578, followed by the unfinished *La Seconde Sepmaine* in 1584. *Les Semaines* were vastly popular in contemporary England and published partial or complete in nine editions between 1592 and 1641 by one translator alone: Josuah Sylvester. Encyclopaedic in their scope, *Les Semaines* looked, amongst other things, to enrich French literature with new images from the sciences, presented characteristically in the form of verbal 'emblems'. In Sylvester's translation 'Geometrie' features as:

That sallow-faç't, sad, stooping *Nymph*, whose eye
Still on the ground is fixed stedfastly,
Seeming to draw with point of silver Wand
Som curious Circles in the sliding sand;
Who weares a Mantle brancht with flowery Buds,
Embost with Gold, trayled with silver Floods,
Bord'red with greenest Trees, and Frindged fine
With richest azure of Seas storm-full brine:
Whose dusky Buskins (old and tatter'd out)
Show she hath trauail'd farre and neere about
By North and South.
   ('The Colvmnes: The IIII Part of the Second Day of
the II Week', Du Bartas 1979, I: ll. 137–47, p. 471)

Du Bartas's 'Geometrie' seems in one respect definitively worldly: she is, in a beautiful figure, dusky-buskined to show her involvement in real, material travel, and her mantle is a worldly map whose elaborate preciousness doubles its materiality. Moreover she is intent upon the very earth itself, rather than abstract philosophy, dealing with a geometry not of the eternal absolute, but of the reeling world: figures in the 'sliding sand' which recall the humanist myth of Egyptian geometric origins and the dusty circles of Joseph's Egyptian mathematicians. Further on in his verses Du Bartas firms up this commitment to a geometry of the world, listing the practical arts, from surveying through commercial navigation to ballistics, that geometry has inspired:

> Heere-by, the wings of favourable Windes
> Shall bear from Westerne to the Easterne *Indes*,
> From *Africa* to *Thule*'s farthest flood
> A House (or rather a whole Towne) of Wood,
> While sitting still, the Pilot shall at ease
> With a short Leaver guide it through the Seas.
> (Du Bartas 1979, I: ll. 217–22, p. 473)

But notwithstanding all this worldliness, Du Bartas's emblematic 'Geometrie' begs certain questions. To look down at the dust in bent-shouldered, scholarly preoccupation is at the very least an ambivalent gesture: does the averted eye of 'geometrie' notice the rich trade and warfare carried out in her name? An ambivalence towards the world is entirely in the spirit of Du Bartas's Protestant poetics, which tempts humanity as Milton would to rational pride, then bids it abandon human reason in favour of faith and revelation. What is not entirely clear, however, is whether geometry is spiritual escape route or worldly trap: whether 'Geometrie' is looking at the world as she contemplates her muddy figures, or away from it. And Du Bartas's geometry is ultimately an ambivalent geometry. It may have dusky buskins, but it can equally seem to breathe its spirit into practical art from a Platonic elsewhere. '*Geometrie*', writes Du Bartas, is:

> The Crafts-mans guide, Mother of *Symetrie*,
> The life of Instruments of rare effect,
> Law of that Law which did the World erect.
> (Du Bartas 1979 I: ll. 158–60, p. 471)

I think the ambiguity of Du Bartas's simultaneously ideal and worldly geometry was entirely characteristic of the early modern cultural currency of geometry and an inherent part of its value, both to mathematicians, and to other professional scholars who traded in the value of mathematics. For Du Bartas, the double identity of geometry functions as a powerful expression of the Protestant humanist's ambivalence towards the world. I regard Du Bartas's 'Geometrie' as a generic form of a figure that appears in various guises in early modern mathematical rhetoric, mediating the unstable meanings and values of mathematical art, and of worldly artfulness in general. This figure will receive further attention in a later chapter. Here I want to focus on just one of its guises: that of Daedalus.

Daedalus's flight

Daedalus's story is told most famously in the eighth book of Ovid's *Metamorphoses*, Englished by Arthur Golding in 1567, and by George Sandys in 1632. Ovid's Daedalus, banished from his native Athens for the murder of his nephew, became a peripatetic engineer. Employed on the island of Crete by King Minos's wife Pasiphae to construct the machine that would allow her to have sex with a bull, he is then called upon by Minos to build a labyrinth to contain her bastard progeny. Sandys's translation describes the commission:

> That uncouth prodigie, halfe man, halfe beast;
> The mothers foule adultery descry'd.
> *Minos* resolves his marriage shame to hide
> In multitude of roomes, perplext, and blind.
> The work t'excelling *Dædalus* assign'd.
> Who sence distracts, and error leads a maze
> Through subtill ambages of sundry wayes.
> . . .
>   so *Dædalus* compil'd
> Innumerable by-waies, which beguild
> The troubled sense; that he who made the same,
> Could scarce retire: so intricate the frame.
>   (Ovid 1970: VIII. 155–61, pp. 357–8;
>   VIII. 166–9, p. 358)

Ovid tells the subsequent story of the labyrinth and then returns to Daedalus, this time in Golding's translation:

> Now in this while gan *Dædalus* a wearinesse to take
> Of liuing like a banisht man and prisoner such a time
> In *Crete*, and longed in his heart to see his natiue Clime.
> But Seas enclosed him as if he had in prison be.
> Then thought he: though both Sea and Land King *Minos* stop fro me.
> I am assurde he cannot stop the Aire and open Skie:
> To make my passage that way then my cunning will I trie.
> Although that *Minos* like a Lord held all the world beside:
> Yet doth the Aire from *Minos* yoke for all men free abide.
> This sed: to uncoth Arts he bent the force of all his wits
> To alter natures course by craft.
>   (Ovid 1961: VIII. 245–55, pp. 164–5)

The craft to which Golding's translation refers constructs the wings with which Daedalus and his son Icarus escapes. Ignoring Daedalus's warning to steer a middle course between sun and sea, Icarus flies too high and is drowned, when the wax which holds the feathers of his wings is melted.

Daedalus, who was careful not to fly too near the sun, is rarely used in discourses of early modern mathematics to signify an unalloyed idealism; this is Icarus's function. Where, as often, Daedalus signifies a pragmatic mathematics, sometimes his pragmatic ingenuity is celebrated as the source of artificial marvels so cunning that they delude some into thinking them magic (see '190: The Prodigall, on himselfe', in Bancroft 1639: sig.E3v). Elsewhere, however, the same ingenious artifice is attacked as malicious worldly cunning (see Du Jon 1638: 124). In the Ovidian translations above, it is interesting that Sandys refers to the Minotaur as 'uncouth prodigie', where in Golding it is Daedalus's arts that are themselves 'vncoth', not reducing nature to an original form, but leading it astray. Golding's Daedalus, it seems, is a Faustian conjuror; Sandys's a Baconian natural scientist.

The ambiguous value of Daedalus's art was manifest to Francis Bacon himself. In *The Wisedome of the Ancients*, Englished in 1619 by Sir Arthur Gorges, and a relative of Du Bartas's compendium of ancient learning, Bacon draws out the significance of Daedalean parables for the more general understanding of 'Mechanicall wisedome and industry' (Bacon 1690: 90). This theme, writes Bacon, 'is shadowed by the Ancients vnder the person of *Dædalus*, a man ingenious, but execrable' (Bacon 1690: 90). Bacon's version of the Daedalus myths emphasizes those elements which he takes to signify 'vnlawfull science peruerted to wrong ends', but also finds in Daedalus's ingenuity the potential for redemption (ibid.: 90). The labyrinth, and in particular Daedalus's indirect aid to Theseus in helping Ariadne provide him with the clew by which he escaped the labyrinth, is fit symbol of this inherent uncertainty in Daedalus's art, being 'a worke for intent and vse most nefarious and wicked, for skill and workmanship famous and excellent' (ibid.: 91). 'The *Labyrinth*,' Bacon concludes, 'is an excellent Allegory, whereby is shadowed the nature of Mechanicall sciences . . . for Mechanicall arts are of ambiguous vse, seruing as well for hurt as for remedy, and they haue in a manner power both to loose and bind themselues' (ibid.: 93–4).

Beyond disjoined celebrations of and satires on ingenious Daedalean art, by far the most characteristic literary use of Daedalus, as the formal characteristics of his most famous story would suggest, is in figurations of the golden mean. The humanist notion of the golden mean is a highly portable one and is often not specifically related to mathematics. Otto Van Veen's *Amorum Emblemata* (1608) uses Daedalus in one of a series of figurations of Love (Figure 3.2). Here Cupid holds a set of compasses and urges we learn that 'Medio Tvtissimus Ibis' (Van Veen 1608: 42).

Elsewhere, Daedalus can be encountered in Renaissance poetry figuring the golden mean in poetic style, social estate and religious devotion (see 'To Dr Scarborough', Cowley 1905–06, I: st. 6, p. 200; 'Medio Tutissimus Ibis',

AMORVM.   43

F 2

Figure 3.2 Otto Van Veen, 'Amorvm', in *Amorum Emblemata, Figuris Aeneis Incisa Studio Othonis Vaeni Batauo-Lugdunensis* (Antwerp, 1996), 43. Reproduction courtesy of the British Library (shelfmark: 96.a.26).

Corbet 1871: 131–2; Ross 1642: 111–14). In Ben Jonson's masque *Pleasure Reconciled to Virtue* (produced 1618; first published 1640), it is the art of life itself that must strive to chart a Daedalean middle way between the virtue of the Platonist's 'unloosed' withdrawal from the world, and the pleasure and profit of 'binding', artful engagement.

Pleasure Reconciled to Virtue enacts a popular Renaissance theme: the choice of Hercules.[9] The narrative of the classical myth sees the youthful hero meeting two women at a crossroads, one named and offering him 'Pleasure'; the other 'Virtue'. Jonson's masque opens with an anti-masque featuring Comus, 'the god of cheer, or the belly', denounced by Hercules as one who wallows 'in the sty / Of Vice' (*Pleasure Reconciled to Virtue*, Jonson 1969: l. 5, p. 263; ll. 93–4, p. 267). Yet Jonson's central thrust is to suggest to his courtly audience, and in particular the young prince Charles, the possibility that whilst pure sensuality must be banished, pleasure and virtue might be reconciled, and that with 'Pleasure the servant, Virtue looking on', worldly appetites might be pursued without fear of 'effeminate' corruption (Jonson 1969: l. 192, p. 271; l. 190, p. 271).

The question begged here is how this principle of good government might be securely established and maintained. Jonson's appeal for an answer is staged and figured in explicitly geographic and even mathematical terms, punning on the artificial mountain which Inigo Jones constructed for the 1618 performance of the masque to the court at Whitehall:

> Ope agèd Atlas, open then thy lap,
> And from thy beamy bosom strike a light,
> That men may read in thy mysterious map
>   All lines
>   And signs
> Of royal education and the right.
>   (Jonson 1969: ll. 196–201, p. 271)

'Through pleasure lead', continues the choir which descends from the mountain singing to Hercules:

> Fear not to follow:
> They who are bred
>   Within the hill
>   Of skill
> May safely tread
>   What path they will.
>   (Jonson 1969: ll. 206–12, p. 271)

Jonson suggests that art, figured in cartographic lines and signs, can help mankind negotiate the pleasures (and by extension, simply the material vagaries) of the world without having to reject them. To ram this suggestively mathematical message home, Jonson's choir are followed in their descent from Atlas by Daedalus, 'A guide that gives them laws / To all their motions' (Jonson 1969: ll. 215–16, p. 272). Daedalus's laws are familiarly Platonic principles of 'sacred harmony', a reasonably common reference point in Jonson's poetry (ibid.: l. 218, p. 272; see Harris *et al.* 1973: 61–2). In humanist style, however, these laws are a means not just of escaping the worldly labyrinth, but of living within and governing it. In Bacon's terms, they both loose and bind.

Daedalus's 'precepts', comments Hercules, 'securely prove / Then any labyrinth, though it be of love' (Jonson 1969: ll. 220–1, p. 272). Stephen Orgel glosses Jonson's 'prove' as 'attempt'. I think the word is best understood as a synonym for the early modern sense of 'improve': 'make the best of' as in (im)proving bread, or the 'proof' measurement of fermented (improved) liquor. Daedalus's precepts enable him to realize out of the 'sty' of thoughtless pleasures in which Comus wallows the disinterested virtue of the scholar's study, just as the baker (im)proves his ingredients to make bread. Like the Egyptian farmer's geometric boundaries, 'discovered' in the featureless river mud, they follow nature, yet refine it.

Rather than explicit metaphors of bakery or brewing, Jonson's court masque prefers the elegant associations of formal dance. 'Come on, come on;' sings Daedalus:

> and where you go,
> So interweave the curious knot,
> As ev'n th'observer scarce may know
> Which lines are Pleasure's and which not.
> ...
> Then, as all actions of mankind
> Are but a labyrinth or maze,
> So let your dances be entwined,
> Yet not perplex men unto gaze;
> But measured, and so numerous too
> As men may read each act you do.
> (Jonson 1969: ll. 224–7, p. 272; ll. 232–7, pp. 272–3)

Life lived in Jonson's courtly labyrinth sees pleasure and virtue so tightly interwoven that they cannot be disentangled. Comus's sty and the scholar's study are balanced in a curious poise, one buried within the other. But although this balanced life is characterized by the playful, superficial ease of dance, beneath this surface, or framing it, are 'precepts' of profound meaning. These are explicitly mathematical: the dance of life is harmonious when it is 'measured, and so numerous', like painting governed by the mathematical principles of design. 'Again yourselves compose', sings Daedalus in a second song:

> And now put all the aptness on
> Of figure, that proportion
> Or colour can disclose.
> That if those silent arts were lost,
> Design and picture, They might boast
> From you a newer ground.
> (Jonson 1969: ll. 249–55, p. 273)

Where Stephen Orgel's classic reading of *Pleasure Reconciled to Virtue* moves quickly through its mathematical and geographic figures to the numbers and forms of 'verse and dance', I think they deserve more independent weight (Orgel 1965: 182). Jonson's masque figures mathematical design as the means for man to refine and live gracefully within a fallen world: to 'improve' it. The dances and verses of the masque appear to me to figure this life of orderly mathematical proportion as much as the other way around. They reflect, after all, not just Jonson's poetic contribution to the masque, but also the contribution of his singularly mathematical partner Inigo Jones, architect, designer and surveyor.

Jones did not just become a mathematical artist, but was highly influential in re-defining mathematical art, much as Dee's preface to Euclid sought to do: as liberal (if useful), and thereby worthy of the social credit associated with the most unworldly scholarship. Jones's Platonist insistence on the ideal grounding of the design-led arts he practised served his anxious purpose 'to achieve parity with the poets who still dominated the cultural scene' (Harris *et al.*: 62). His success in promoting such an image of himself as pragmatic Platonist can be judged in his commissions: Jones was the first English architect to be given sole responsibility for the design of major building projects and much of the credit for their accomplishment (ibid.: 25). It can also, perhaps, be judged in a masque whose Daedalus might be either Jones or Jonson, leading the court from chaotic anti-masque to orderly masque.

Daedalus figures once again, if only by unmistakable implication, in another general account of human arts, which once again generates an intermediate mathematics as first principle of a virtuous 'improvement'. Fulke Greville's long verse treatise *A Treatie of Humane Learning* considers the scope and value of human knowledge. Greville's poem, never published in his lifetime, is thought by his editor Geoffrey Bullough to be in part a response to Bacon's *Advancement of Learning* (1605), but may not have been composed until after Greville's loss of public office in 1621 (Bullough 1939: 17, 52; Gouws 2004). As Bullough comments, it is Senecan and Calvinist until its sixtieth stanza, rejecting human knowledge with all the vehemence of Protestant complaint and the scepticism that would continue to dog advocates of scientific curiosity throughout the Stuart era (Bullough 1939: 60; see Webster 1975: 17). From thereon, however, it takes a Baconian turn, embracing the practical arts with what Bullough takes to be an ironic, Machiavellian utilitarianism as the best means for the un-elect, at least, to navigate a fallen world (Bullough 1939: 13). Matthew Woodcock notes that Greville's five long verse treatises all follow the same pattern: beginning with a prelapsarian perfection forever lost, but then constituting human art as capable of re-'moulding' man and nature (Woodcock 2001: 148).

In his early stanzas Greville rehearses Plato's hierarchy of human comprehension from the intellect through imagination to the senses, and grants that human 'faculties of apprehension' may, as Plato suggests, have been perfect 'in the soules creation' (Greville 1939, I: st. 18, p. 158). But unlike Plato he despairs, as a Calvinist, of the capacity of even the highest human reason to free itself from the stain of mankind's fallen nature. Liberal sciences and arts he deems a vain attempt to redeem this stain (ibid.: st. 21, p. 159). If indeed there are 'generall, vniforme Axioms scientificall / Of truth, that want beginning, haue no end' this proves those arts that contain them 'proper to the *Deity*' (ibid.: st. 22, p. 162). In the stead of those Platonic 'characteristicall *Ideas* ... which Science of the Godhead be' humanity has established arts that, whatever their pretentions to the absolute, have more of earth about them than eternity (ibid.: st. 25, p. 163).

Greville's notion of an unobtainable geometric ideality allows him to associate geometry both with its fallen, pragmatic status, and the Platonism to which it still vainly aspires. Both make geometry the butt of his scorn. In stanzas 32 and 33 of his poem Greville introduces geometry in Ramist fashion as the humble science which 'gives measure to the earth below', reprimanding such materialism with what we can recognize as the conventional Platonisms of Protestant devotion:

> Rather let her instruct me, how to measure
> What is enough for need, what fit for pleasure.
>
> Shee teacheth, how to lose nought in my bounds,
> And I would learne with ioy to lose them all.
> (Greville 1939, I: st. 32–3, p. 162)

Notwithstanding these stock aversions from the worldliness of geometric practice, and specifically the surveying of property, it is ultimately for the Platonist geometer that Greville reserves greatest contempt, and here he expresses the Reformation-humanist philosophy of geometry with passion. 'The grace, and disgrace' of geometry, he writes:

> Rests in the *Artisans* industrie, or veine,
> Not in the Whole, the Parts, or Symmetrie:
>   Which being onely Number, Measure, Time,
>   All following Nature, helpe her to refine.
> (Greville 1939, I: st. 116, p. 183)

Here explicitly is the Ramist and Baconian closed circuit of a theory inductively drawn from use, and intended to inform and better further use. Not, Greville emphasizes, to inspire a Platonic, Icarean ascent up the ladder from sense through imagination to philosophical contemplation. 'Practise in materiall things' should not 'awake that dreaming vaine abuse / Of *Lines*, without *breadth*; without feathers, wings', but be bounded 'In Workes, and Arts of our Humanity' (ibid.: st. 118, p. 183). Too often those who dabble with mathematical arts are:

> So melted, and transported into these;
> And with the Abstract swallowed up so farre
> As they lose trafficke, comfort, vse, and ease
> (Greville 1939, I: st. 119, p. 183)

The only reformation of these arts worth attempting, concludes Greville, must be:

> By carrying on the vigor of them all,
> Through each profession of Humanity,

Military, and *mysteries Mechanicall*:
> Whereby their abstract formes yet atomis'd,
> May be embodied; and by doing pris'd.

As for example; Buildings of all kinds;
Ships, Houses, Halls, for humane policy;
Camps, Bulwarkes, Forts, all instruments of Warre;
Surueying, Nauigation, Husbandry,
> Trafficke, Exchange, Accompts, & all such other,
> As, like good children, do aduance their mother.

For thus, these Arts passe, whence they came, to life,
Circle not round in selfe-imagination,
Begetting *Lines* upon an abstract wife,
As children borne for idle contemplation;
> But in the practise of mans wisedome giue,
> Meanes, for the Worlds inhabitants to lieu.
>   (Greville 1939, I: st. 122, p. 184)

In Greville, as in Du Bartas, we find a fascinating ambivalence about the origins and value of geometry. Geometry here would seem to have an ideal basis which yet must be ignored or atomized. It would equally, if paradoxically, seem to come from and return to human 'life', a proper, more valuable circuit than the sterile one of idle, abstract contemplation, and yet such worldliness, if preferable to philosophy, can also seem too worldly. Greville averts his own eyes from the ideal and recommends we do the same, embracing worldly practices for which he and we must yet feel a certain distaste. The effect of this ambivalent commitment to practice might seem to be to remind us of the ideal principle that lies behind the mud-drawn geometric line and compensates us for its dispiriting corporeality. Most crucially of all, for this book, it informs a notion of art as 'natural refinement': the civil improvement of human life through principles derived paradoxically from human life itself.

Greville articulates what Matthew Woodcock calls 'an enduring belief in the capacity of correctly ordered temporal institutions and faculties, combined with faith, to frame or mould mankind in order that we might play a positive, active role in the postlapsarian world' (Woodcock 2001: 159). And the equivocal status of Greville's mathematics – part of the world; part beyond it – uniquely suits it to form such a frame. Mathematics functions for Greville, as for Du Bartas, to express an inherent humanist ambivalence intensified by Calvinism. Most importantly it is thereby constructed as the means to negotiate such ambivalence: a means to live in the world whilst remaining distant from it. Charles Webster suggests that the Baconian scientist negotiated the scepticism of his contemporaries by limiting his science to worldly use (Webster 1975: 22). Greville's ambivalent mathematics suggests a more complex negotiation between utility and ideal value.

In another of Fulke Greville's verse treatises the politics of this ambivalent, improving mathematics are made explicit, in a mercantilist theorisation of economic value. In the ninth book of his *A Treatise of Monarchy* (composed *c.*1610), titled 'Commerce', Greville argues for the benefits to the nation of developing and maintaining trade (*A Treatise of Monarchy*, Greville 1965: IX). Here the artful sense of improvement is much at stake. Kings should not, Greville urges, see competition in the self-advancement of the skilful artisan and tradesman, but should do everything possible to encourage their activities as fostering the common wealth:

> Wherefore with curious prospect theis prowde Kings
> Ought to survey the commerce of their lande;
> New trades and staples still establishinge,
> So to improve the worcke of everie hand,
>   As each may thrive, and by exchange, the throne
>   Growe rich indeede, because not rich alone.
>   (Greville 1965: IX. 378, p. 129)

Hard work and labour here are contributions to the common stock. Most importantly, it isn't simply brute hard work that Greville advocates, but art: the improvement of raw materials. Greville argues that the English must do as the Hollanders do, and:

>   worcke her matter with her home-borne hands,
> And to that use fetch forraine matters too,
> Buyeinge for toyes the wealth of other Lands,
> . . .
>
> Wherein wise Princes ought to imitate
> The *Saracens* inriching-industry,
> Who *Egypts* wealth brought to their barren state,
> Entisinge vyce by farr fetcht vanitie,
>   And for their Ostridge feathers, toys of pride,
>   Gett staple wealth from all the worlde beside.
>   (Greville 1965: IX. 392–3, p. 133)

Greville is far from able to disassociate from improvement the taint of luxury; of the unnecessary and effeminate toys upon which prodigal sons waste their father's wealth. However, as long as the traffic in such toys is one-way, and its return is made in solid useful commodities, and best of all bullion, Fulke Greville sees no harm in improvement. Greville finds in 'art' itself the most certain guarantee that the products of artful commerce will not corrupt. The core of such artistic virtue he figures in explicitly mathematical terms:

> Yet must there be a kynde of faith preserv'd
> Even in the commerce of the vanitie,

That with true arts their marketts may be serv'd,
And creditt kept to keape them greate, and free;
   Weight, number, measure trulie joyn'd in one,
  By Trade with all states, to inrich our owne.
             (Greville 1965: IX. 388, p. 132)

Mathematics here stands figurative guarantee of a general improvement that constrains the potential threat of luxurious and rebellious individualism by generalizing its benefits. 'In States well tempered to be rich', writes Greville:

Arts be the men's, and men the Prince's are;
Forme, matter, trade so worckinge everie where,
As governement may finde her riches there.
             (Greville 1965: IX. 389, p. 132)

Greville's evocation of a harmonious mathematical economy is markedly reminiscent of Ben Jonson's improving Daedalean dance. Like Jonson advising his prince on the reconciliation of virtue and pleasure, Greville clearly feels that the individual benefits of improvement require some frame to guarantee their incorporation within a general economy of virtue: to 'temper' them. He invokes practical mathematics to figure this tempering frame, urging proud kings to 'survey' a realm of burgeoning art, commerce and general improvement.[10] Made confident by their mathematical overview, these kings should not fear that such a realm will exceed the due bounds of virtuous government.

Greville's is an anxious argument, hedged around by anxieties about the possibility of reconciling virtue with commerce. Moreover his confidence in mathematics as 'a kind of faith preserv'd / Even in the Commerce of the vanity' is undercut by the scorn he expresses elsewhere for both sterile mathematical Platonism and venal mathematical materialism. Greville's mathematics, as Francis Bacon so aptly put it, both 'binds' and 'looses': it is of an equivocal, Daedalean value. This ambivalence is not simply a rhetorical fissure, however; the product of painful personal tussles between Calvinism and Baconian humanism. It is also a powerful rhetoric of negotiation to which both of these cross-fertilizing traditions subscribe.

4 Discipline reconsidered

A perspective to look beyond tradition

Fulke Greville's poetry suggests a reconsideration of the role played by mathematics in a Protestant culture of discipline. It suggests that if mathematics was entangled with an emerging capitalist ethos it was as an equivocal rhetoric of balance and constraint, rather than an instrument of confident dominance, exploitation and control. In the following chapter I want to return to the discourses and aesthetics of a Protestant culture of discipline, and to consider the role played by an equivocal mathematics in negotiating the controversial values of seventeenth-century economic reform.

Girdling the English waste

Previous studies of early modern discipline have made much of its gendered nature (Kolodny 1975; Montrose 1991). The figurative equivalence of woman with land is a well-known early modern trope which worked both ways round. Most striking and persistent amongst all the terms of a 'landscaped' sexual economy is the figuration of unbridled feminine sexuality as waste and common land, and vice versa. Gervase Markham, a prolific writer better remembered for his agricultural than his poetic publications, places this figure in the mouth of Paulina, in his 1609 dramatic monologue *The Famous Whore, or Noble Curtizan* (Markham 1609). Paulina describes the difficulties of her churchman lover in monopolizing her affections, 'Knowing that hard it was of common ground, / To make a priuate walke, or so inclose it, / As law, or scandall would not make him lose it' (ibid.: sig.B2v).

Markham acknowledges the controversial nature of enclosure here: opposed not just by the general 'scandall' of the populace, but also by a body of law still infused with the values of custom and stewardship. This sense of the 'scandall' of enclosure is reinforced by the identification of the encloser with the luxurious, lecherous priest. At the same time, 'common ground' is hardly the virtuous term opposed to the encloser's vice, identified as it is by Paulina herself with her prostitution. Elsewhere the prostitute and unenclosed ground are made mutual targets of a less ambiguous moral indictment. In John Taylor's 1630 poem 'A whore' the prostitute's girdle 'incloses' not a deer park, but a 'wastfull waste', and Alexander Garden's

'Baude', number 41 in his collection *Characters and Essays* (1625) also has a 'waste Wombe' ('A whore', John Taylor 1630: 111; 'A baude', Garden 1625: 50). Shakespeare's Angelo in *Measure for Measure* finds in a more general feminine wasteness the opposite of chaste Isabella's virtue:

> Can it be,
> That modesty may more betray our sense
> Than womans lightness?
> Having waste ground enough,
> Shall we desire to raze the sanctuary
> And pitch our evils there?
> (*Measure for Measure*, Shakespeare 1986: II. ii. 173–7, p. 798)

If the outer limit of feminine licence for a Puritanical misogyny is the blasted waste-ness of the prostitute's barren womb, a formal aesthetic of enclosure is widely used in early seventeenth-century writing to denote the containment of human sexuality, conventionally coded feminine. A common figure of simultaneously physical and legal constraint in seventeenth-century writing is that of the girdle. Canto III of Joseph Beaumont's *Psyche* (1648), titled 'The Girdle', uses enclosure to figure the laws enjoined upon fallen man, and specifically those constraining sex. Here Herod is reminded by John the Baptist of the prohibition against love between men and women related by marriage:

> *God* who made this enclosure, hedging Her
> In to her *Philip*, still hath left to Thee
> And thy free choice, an open Champain, where
> Millions of sweet and virgin Beauties be.
>   Adorn thy bed with any one beside,
>   Only *thy Brother's must not be thy Bride*.
> (*Psyche, or, Love's Mystery*, Beaumont 1880, I: III. 160, p. 55)

Edmund Waller, in 'On a girdle' (1645), celebrates the same, now markedly mathematical constraint of land and woman:

> It was my heaven's extremest sphere,
> The pale which held that lovely deer,
> My joy, my grief, my hope, my love,
> Did all within this circle move!
>   ('On a girdle', Waller 1893, I: 95)

In a 1653 elegy by William Basse it is the speaker's own imposition of mathematical constraint that charms him: 'on the easie measure of hir waste / I in this sort desiringly fell mad' ('Elegie III', Basse 1893: 84).

The terms of these poems – of an open 'waste' of feminine sexuality enclosed in a hedge, girdle or measure of marriage or individual love – are remarkably persistent. Even where landscaped sexualities appear to celebrate sexual freedom; a kind of pastoral sexuality, they mark enlosure as the natural state for fallen man.[1] Where waste land can be viewed not as a real and concrete resource for the rural poor, but as the physical topos associated with a lost state of innocent liberty, enclosure can be characterized as the inevitable advent of improving constraint; of 'measure' in a corrupt and fallen world. In 'Upon Appleton House' (1681) Marvell imagines the tenants of the nunnery from which Fairfax's mansion was developed:

> Within this holy leisure we
> Live innocently, as you see.
> These walls restrain the world without,
> But hedge our Liberty about.
> These bars inclose that wider den
> Of those wild creatures, callèd men.
> The cloyster outward shuts its gates,
> And, from us, locks on them the grates.
> ('Upon Appleton House', Marvell 1972: ll. 97–104, p. 78)

Demonstrating the ambivalence for which Marvell is famous, these figures play on the ambiguity of the constraint that also liberates; the binding that somehow also looses.

In those mid seventeenth-century agrarian treatises which explicitly advance the agenda of the improver the figures of a gendered disciplinary discourse are frequently used to disparage customary commonality and to naturalize enclosure as the necessary government of fallen man. In *Bread for the Poor* (1653) Adam Moore speaks of the commons as a 'common prostitute', through which the commoner is 'cuckolded by Forreigners and strangers' (Adam Moore 1653: sig.A2r). 'Your Common,' he mocks, is 'used before your face, even as commonly as by yourselves' (ibid.: sig.A2r). Moore deploys the strongest 'Georgic' rhetoric on female promiscuity to castigate common land, ascribing it physical qualities of ugliness and barrenness conventionally associated in contemporary discourse with the prostitute (see Laqueur 1990: 230). For Moore, commons are cursed with the void-ness of the common woman's empty womb: they are 'gulfes of want and penury in a deformed visage'; they are 'fruitlesse, naked, and desolate'; they are, '(as nature in defect) ... delivered of nothing but Monsters' (Adam Moore 1653: 17, sig.B1V, 20). In another enclosure polemic the sexualized disorder of the unbridled commons is associated with the social radicalism of those 'levellers' who see common rights as the basis for a 'liberty of conscience' (Lee 1656: 29). Such 'levellers' desire, thinks Joseph Lee:

> to live as the [*sic*] list and sin *cum priveligio*; the golden reins of discipline please not, this yoke they cannot bear, but cast off this, and then they

may swear, and lie, and rob, and rifle, and swill, and swagger, riot and revell in a shorelesse excesse ... The very same spirit of disorder in the very same persons, and upon the same grounds doth decry Inclosure because it would put a bridle upon their licentious lusts.

(Lee 1656: 29–30)

Where the traditional moral economy conceives of a national polity constituted through the aggregate of its particular local traditions, the formalized economy of enclosure polemic makes common use seem incompatible with any wider human intercourse, including that of language itself. Silvanus Taylor, a minister from Cotesbach in Leicestershire whose tract was a response to violent attacks from his peers for the part he took in the enclosure of 'Catthorp Common', complains that the very name of *'Commons'* is a 'cloud generated to darken all that have not a Perspective to look beyond Tradition' (Prothero 1961: 125–6; Silvanus Taylor 1652: 11). Common land in enclosure polemic cries out for a Georgic 'girdle'; for 'golden reins of discipline' to bridle in its 'shorelesse excesse'; for a 'Perspective to look beyond Tradition'. In a tract authored by Cressey Dymock, and published as a letter from Dymock to Samuel Hartlib in 1653, the girdle is put in place.

Dymock's *A Discovery for New Divisions; or, Setting out of Lands, as to the Best Forme* projects a design whose regular form reduces the complexity of the post-feudal manor to the simplicity of universal geometry and unmediated private property (Figure 4.1). Dymock instructs his reader to draw 'a Circle from your House, at the Centre as wide as your Square will admit', and, elaborating on his concentric lines of demarcation, 'to binde all this together ... encompasse all those with one undivided ring' (Dymock 1653: 19, 20). 'Finally,' Dymock concludes, 'here your house stands in the middle of your little world ... enclosed with the Gardens and Orchards ... that again encompast with little Closes ... all bound together as with a girdle' (ibid.: 10). Dymock's ideal farm or reformed manor is bound together; 'girdled' like a well-kept Georgic wife. But through its overlapping rhetorics of gender and geometry it turns its Georgic limits outwards, permitting Dymock's editor, Samuel Hartlib, to depict it as a 'Publique Design' involved in laying 'the foundations of Trade and Commerce': part of a 'Reformation' of 'Disorderliness' and 'Confusion' into 'righteous Order' (Hartlib 1653: sig.A2r, sig.A2v).

Equal balance and wondrous gain: disciplining improvement

What is bound and loosed in the gendered discourse I've described so far is a feminized landscape/landscaped femininity which cannot be left common or waste in a lapsarian condition without fear of corruption. But equally importantly, as it works to valorize a Georgic aesthetic of enclosure, this dis-

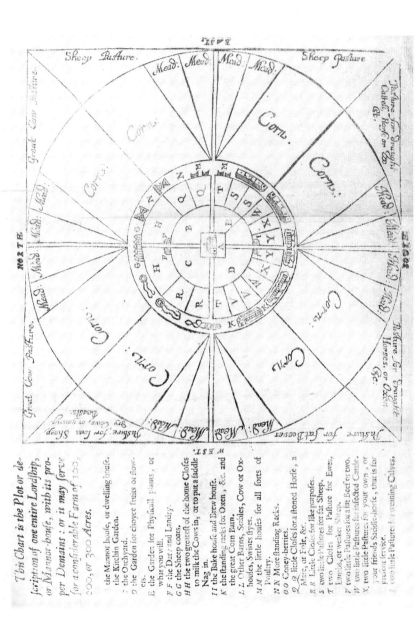

Figure 4.1 Cressey Dymock, 'This Chart is the Plot or description of one entire Lordship, or Mannor-house, with its proper Demains; or it may serve for a considerable Farm of 100, 200, or 300 Acres', in Samuel Hartlib (ed.) *A Discoverie For Division Or Setting out of Land, as to the Best Form . . . whereunto are added some other choice secrets or experiments of husbandry* (London, 1653). Reproduction courtesy of the British Library (shelfmark: 234.e.32(1)).

course works simultaneously to counter the traditional association of improvement itself with wasteful, effeminate, luxurious corruption. If the female body is the test of an improving discipline brought to bear on the promiscuous commonality of custom, it also tests the corrupting tendencies inherent in improvement itself.

Associations with wanton luxury haunted the notion of improvement throughout and beyond the seventeenth century. 'By this light,' exclaims Nashorat, in Sir John Suckling's 1646 comedy *The Goblins*:

> I hate a woman drest up to her height,
> Worse then I doe Sugar with Muskadine:
> It leaves no roome for me to imagine:
> I could improve her if she were mine.
> (*The Goblins*, Suckling 1648: IV, p. 42)

Whether intentionally or not, Suckling's lines suggest a bawdy double entendre: that of the (im)proved and thereby swelling womb that is the proof of sexual licence. This pun on the 'improvement' of women is more explicitly illustrated in Robert Heath's epigram 'On Galla her going to a nunnerie' (1650):

> E'r her Probation year was finished,
>   She not approv'd that life; *Improve* she did:
> The first year *Galla* only said she meant
> To prove: She prov'd indeed, with child, and went.
>   ('On Galla her going to a nunnerie', *Epigrams*, Heath 1650: 3)

Heath's puns suggest the close association of improvement with an 'approval', or legitimacy founded on the testing and realization of potential. However, they also suggest an anxiety about the proper containment of such development expressed in sexualized and gendered terms: the terms of female promiscuity. Countering these dangerous associations in seventeenth-century discourse are figurations of surveying and enclosure as the Georgic containment; the reduction to good measure of excessive desire: the desire simultaneously of the bawdy commons and, in a mighty paradox, of the rapacious encloser himself.

A discourse and aesthetic of formal containment offers an alternative model for the disciplining of human desire and market forces to the older paradigm of patriarchal stewardship, which historians have suggested had entered a period of crisis by the early seventeenth century. Whilst the professional role of stewardship expanded with the market economy, literary representations of stewards tend to associate them closely with 'the broader sense of the term' stewardship: the preservation of the traditional moral economy of the manor (Hainsworth 1992; Sullivan 1998: 160). These literary representations seem to respond to a perceived need for reassurance that the old patriarchal model still obtains. They can also seem to negotiate between traditional and new socio-economic values. Representations of stewardship on the early modern stage are certainly, as Garrett Sullivan has suggested, a powerful means of embracing a newly managerial approach to land within a reassuringly stable, patriarchal vision of the socio-economic order (Sullivan 1998). But the stage steward is persistently figured as an anachronism who either fails in his bid to girdle the luxurious excesses of a market culture, or mutates into the more appropriately modern figure of the surveyor.

Richard Brome's *A Jovial Crew, Or, The Merry Beggars* (produced 1641; first published 1652) begins by establishing the relationship between Oldrents, an ancient esquire, and Springlove, his steward (Brome 1968).[2] Oldrents is the model of good feudal paternalism, of which his parasitical friend Hearty reminds him. Has he not, Hearty asks, the independence of Fortune secured by four thousand a year, and 'the praises of the rich, / And prayers of the poor?' (Brome 1968: I. i. 68–9, p. 18). The gratitude of poor tenants, in particular, Hearty associates with Oldrents's generosity as a land-lord. 'Whose rent did ever you exact?' he asks his friend; 'whose losses / have you not piously repair'd?'; 'What hariots have you tane from forlorn widows? / What acre of your thousands have you rack'd?' (ibid.: I. i. 80, p. 19; I. i. 83–4, p. 19; I. i. 85–6, p. 19).

Oldrents's perfection as a landlord is almost matched by the perfection of Springlove as a steward, but for the wanderlust which annually drives him to seek the company alluded to in the play's title. The first scene of the play sees the two characters doing their 'accompts'. Turning over the 'several books' to his master, Springlove glosses their contents, concluding with the 'large Benevolences' he has distributed to the poor. Oldrents responds approvingly:

> Thy charity there goes hand in hand with mine.
> And, Springlove, I commend it in thee, that
> So young in years art grown so ripe in goodness.
> May their Heaven-piercing prayers bring on thee
> Equal rewards with me.
> (Brome 1968: I. i. 139–40, p. 21)

Springlove's final totting up, giving a total of 'Twelve thousand and odd pounds' to deposit in his master's coffers, makes this assurance of equal (spiritual) reward appear rather ironic, but nonetheless so harmonious is Springlove's self-incorporation within the economy of Oldrents's estate, that Oldrents thinks his steward's closet as safe a depository of this money as his own (ibid.: I. i. 147, p. 22).

In the figure of Springlove, prudence seems perfectly harmonious with that other essential stewardly quality of honesty. As one whose fault is not to desire his own material advancement, but the free pleasures of beggary,

Springlove is comically immune to the Judas taint of ambition and treachery. John Hall's prefatory eulogy to Brome identifies him as a disciple of Ben Jonson, and concludes: 'So that we must in this your labor find / Some image and fair relic of his mind' ('To Master Richard Brome', Brome 1968: 5). Indeed *The Jovial Crew* might well be regarded as doing much the same ideological work as Jonson's *To Penshurst* is often judged to do, constructing a feudal economy which runs on generosity, loyalty and goodwill, and within which the pleasures of the rich pose no threat to the well-being and gratitude of the poor. Within this economy the steward's executive role as distributor of 'gifts' and collector of rents and profits is in complete harmony with the interests of both landlord and tenant, his honesty guaranteed, finally, by a binding contract of Christian faith. An idealized version indeed of Brome's relationship with Jonson as his 'literary factotum' and go-between in negotiations between Jonson and his actors (Butler 2004).

And yet there are worrying gaps in this account. Oldrents's merry generosity is excessive, as Garrett Sullivan's extensive and penetrating analysis of this play points out, figuring not just as the glue that binds together his estate, but as a disruptive agency threatening to waste it (Sullivan 1998: 186). And in any case the generous Oldrents is clearly an exceptional case, to be regarded, regrettably, as an anachronism in what Brome refers, in his dedicatory epistle to his patron, as 'this Cuckoo time' of insincerity and self-advancement ('To the Right Noble, Ingenious, and Judicious Gentleman, Thomas Stanley, sq.', Brome 1968: 4). No less exceptional is Springlove, who turns out to be no mere hired hand after all, but Oldrents's lost son. Brome's play performs, as Sullivan observes, an 'ironic reversal' of a reality wherein it was the landlord that was most likely to renounce his duties to go a-wandering, and the steward who was most likely to be left behind in charge (Sullivan 1998: 172). It is a self-consciously 'nostalgic ... vision of a paternalistic order based on the estate' (ibid.: 185). Oldrents and Springlove might be compared, in this regard, to another more famous pairing of lord and 'steward': Shakespeare's Lear and Kent. An influential Marxist line of interpretation on Shakespeare's play sees it pitting feudal values and social constructs of custom, loyalty and household against those of an incipient capitalist individualism personified in Edmund (Kettle 1988: 70–8). Whilst the feudal values are indubitably the virtuous ones in the play, they are also made to seem naive to the point of blindness, and doomed in the face of superior creative energies.

Edmund's accomplices in the challenge he poses to the old feudal order are of course sexually predatory women with whom he practises adulterous licence. This gendering of the waste-ful threat to a post-feudal economic order is a common topos in Jacobean and Stuart satire, often countered by the figure of stewardship. In James Shirley's comedy *The Lady of Pleasure* (produced 1635; first published 1637) the lady of the title is Aretina, wife of Sir Thomas Bornwell, a man who complains to his wife in the opening scene:

> Have I not obey'd
> All thy desires, against mine owne opinion,
> Quitted the countrie, and remov'd the hope
> Of our return by sale of that fair lordship
> We liv'd in, chang'd a calme and retir'd life
> For this wild town, compos'd of noise and charge?
> (Shirley 1962: I. i, p. 6)

Released from the girdle of her feudal responsibilities, Aretina enjoys and bestows the chaotically unregulated hospitalities of the town. Accounting for his wife's extravagances much as Brome's steward gives an introductory survey of Oldrents's estate, Bornewell complains to her that her 'wayes of pride and costly ceremony' threaten:

> To stifle us at home, and shew abroad
> More motley than the French or the Venetian
> About your coach, whose rude postilion
> Must pester every narrow lane till passengers
> And tradesmen curse your choaking up their stalls,
> And common cries pursue your ladyship
> For hind'ring o' their market.
> (Shirley 1962: I. i, p. 7)

Bornewell does not stop at his accusation of material incontinence in the 'narrow lanes' of London, hinting at sexual misdemeanours which consume more 'fame' than 'purse' (ibid.: I. i, p. 8).

The association of material luxury with femininity and the urban, and with a simultaneously economic and moral corruption, is a convention of that tradition of political discourse whose roots can be traced to Plato's *Republic*. What is interesting in the present context, however, is not just the conventional pitting of effeminate urban luxury against independent country virtue, but the use of the steward as spokesman for the latter. Aretina is 'Brought in the balance', as she complains, not just by her husband, but first of all, and in the opening exchanges of the play, by her steward (Shirley 1962: I. i, p. 7). Where Aretina celebrates her escape from the crude rusticity of the Lordship, the steward recalls:

> You liv'd there,
> Secure and innocent, belov'd of all,
> Prais'd for your hospitality, and pray'd for.
> (Shirley 1962: I. i, p. 5)

Moreover, this moral pairing of lady and steward is doubled in the figures of another lady of the town, Celestina, and her steward. This steward berates his lady's extravagance with:

> 'Tis not for
> My profit that I manage your estate
> And save expense, but for your honour, madam.
>> (Shirley 1962: I. ii, p. 7)

Small thanks he gets. This moral economy, like Brome's, is coming apart at the seams. The honesty and diligence of the steward, valiant as he is in striving to keep these seams together, and to contain his mistress's burgeoning waste, is scarcely adequate to the task.

Wasteful women on the early modern stage, and those modern, commercial aspects of early modern culture coded feminine, are the enemies not just of a traditional moral economy of duty and hospitality, but also of newer values of Georgic thrift and independence. For even a limited expansion of a commercial economy to be reconciled with moral virtue, the association of commerce with the unfettered burgeoning of effeminate luxury had to be warded off: tempered by some limiting principle of virtue. Whilst he often features as the guardian of the traditional moral economy, modified in modern times by new Georgic values of thrift and prudent management, the steward can often seem inadequate to this task; impotent, like Shakespeare's Kent, in the face of superior creative energies. The 'crisis of aristocratic identity' that Garrett Sullivan finds registered in Brome's *The Jovial Crew* is simultaneously a crisis of confidence in the steward's ability to contain such energies, generated both within and without the boundaries of the manor (Sullivan 1998: 186). Whilst the social practice of stewardship may have burgeoned in the seventeenth century, in ideological terms it was supplanted or at least shored up by the mathematical discourse and aesthetic of surveying. The successful steward begins to seem more and more like a surveyor, and never more so than in the case of Robert Aylett's Joseph, the paragon of seventeenth-century stewards.

Aylett's *Ioseph, Or, Pharaoh's Favourite*, a poetic expansion of scripture first published in 1623, is a text utterly preoccupied with mediation, written by a 'loyal lieutenant' of Archbishop Laud (Steggle 2004). On its title page it bears the legend from Ecclesiastes: 'Hee only that applyeth his minde to the Law of the most high, and is occupied in the Mediation thereof; shall serue among Great men and appeare before the Prince' (Aylett 1623: n.p.). *Ioseph* is dedicated to '*Iohn* Lord Bishop of *Lincolne*, Lord Keeper of the *Great Seale*'. The life of this great 'seruant' and the life of Joseph are '*but one narration*', writes Aylett, both being entrusted with the 'equall Balance' of a nation. And yet under the guise of such disinterested moral stewardship, the mediation of divine law, Aylett uses Joseph to suggest a different, more pragmatic kind of stewardship: one much more concerned with material and individual benefit. Although he explicitly encourages the bishop to imagine himself as Joseph, offering Godly counsel to the King, Aylett's poem positions itself and its author in a further, more equivocal service relationship to this patron, balancing virtuous counsel with worldly service.

In Aylett's poem Joseph's first Egyptian master Potiphar:

> Had in his youth a skilfull Merchant beene,
> And *Stewarded* so frugally his owne,
> That *Pharoh* wise, to whom all this was knowne,
> Mad him chiefe steward; they that can hold fast
> Their *owne*, their *Masters* treasure seldome wast.
>
> (Aylett 1623: 50)

Potiphar came to buy the enslaved Joseph in the exercise of these wisely selfish virtues, buying goods for Pharaoh in the market at first, rather than 'second hand', for 'Thrift stands not on nice court-like superstition' (Aylett 1623: 50). As Potiphar's steward, Joseph manages the great man's stock according to the same Georgic values of prudence and thrift, maintaining his resource. But as Pharaoh's man his role is altogether more radical. Telling his story to his father Jacob, Joseph describes his strategy to accumulate for Pharaoh's Egypt a surplus to set against the famine he had foreseen:

> No Corne out of the Land let I goe out,
> But buy in rather from the Coasts about,
> And many *Forrests* which before did ly
> All waste, I vnto *Tillage* did apply:
> Thus I proceede, and God so blest my hand,
> That all things prosper ouer all the Land.
> But when the yeares of plentie all are past,
> And all the Land of *Egypt* lyeth waste,
> So that they liue of former yeares remaines,
> Which them perhaps a month or two sustaines,
> The people first of *Pharoh* seeke supply,
> Who them to *Ioseph* sends; I suddenly
> Set ope the Barnes, and sell for money out
> The Corne to all the Nations round about.
>
> (Aylett 1623: 61–2)

There are two uses of waste here, distinctly different in their meanings. The second describes the effects of devastation – the wasting of the land by famine. The first, however, refers to land which is simply neglected: surplus to the Egyptian economy of agrarian production. In effect Joseph has decided to respond to the exigencies of his time – a shortage of corn – by becoming an improver, enclosing and developing hitherto neglected land. That Aylett has chosen to speak of 'forests' in this context, with reference to a land which had none, intensifies this distinctly local and contemporary spin on the story. Moreover, what follows does so still more strongly. We have seen that Joseph resorts to a form of enclosure and improvement in order to maintain his master's and his nation's resources. In other words, although he adopts capitalist methods, these are licensed ultimately by the values of moral, or at least Georgic, stewardship: the maintenance, rather than expansion of individual stock. However, this not all that Joseph does. Once he has 'engrossed' the market in corn on Pharaoh's behalf, he announces his intention to exploit this position by a more radical intrusion on the existing socio-economic order:

> I will for Corne their Bodies buy and *Land*,
> But all for *Pharoh*: whereby growing strong,
> He and his Empire may continue long:
> And to establish more his Segnurie,
> From *Place* to *Place* I will each *Colonie*
> Transplant, who on the *East* of *Nile* abide,
> I will remoue vnto the other side:
> Thus *Pharoh Lord of Egypt* shall be knowne,
> By *Seisin*, none shall say this is mine owne.
> Thus haue I seene when *Want* or *Waste* compell
> A *Gallant* his Inheritance to sell,
> (Lest any right in him should still be thought)
> Giue place to him that hath the purchase bought,
> And in another Countrie dwelling hire,
> Whether he with his houshold may retire.
> Thus will I doe with all the *Common Lands*
> (Aylett 1623: 62–3)

Aylett allows Joseph to express the wish-fulfilment fantasy of the English encloser and indeed colonist. Although he invokes the luxurious wasting of the prodigal to license the enclosure of bankrupt estates and the 'engrossing' not just of waste forest but of rich Nile delta land, it is the households of established communities that Joseph wants to dispossess of this fertile resource, turning 'Common Lands' into private enclosures. Moreover, Aylett's Joseph goes still further in his progress from obedient feudal steward to artful capitalist improver. In a remarkable flight of scriptural elaboration Aylett has his hero survey and project the 'channel' which would ultimately be achieved as the Suez Canal:

> Now, as I said, when I did Circuit ride,
> And *Egypts* Land suruai'd from side to side:
> One thing of greatest vse I did obserue,
> . . .
> A peece of Land an *Istmos*, *Barre*, or *Stay*,
> Twixt *Midland Sea*, and the *Arabian* Bay,
> Suppos'd some (a) ten Miles ouer at the most,
> Asioineth fast vnto th'*Egyptian* Coast

Which if one by a *Channel* did diuide,
Both Seas might each into the other slide.
This if it were once nauigable made,
Would bring, to Kings and people of each trade,
Such wondrous gaine as cannot be expected,
With endlesse fame to those which it effected:
...

 This *proiect* I to *Pharoh* did commend,
Intreating him this *princely worke* t'intend,
(Aylett 1623: 63–4)

In Aylett's *Ioseph* the figure of stewardship serves to mediate between the most conservative and the most progressive conception of land ownership and management: the discourse of surveying and projection. Framing Pharaoh's economy, just as Aylett's patron secures the 'equall Balance' of the state, Joseph stuffs this frame with 'wondrous gaine'.

Aylett's poem tugs at the sleeve of a potential patron, projecting schemes of material improvement which Aylett seeks nonetheless to encompass within a girdle of 'equall balance' and godly stewardship. If surveying is tangled up with improvement here it is not simply as the pragmatic tool of a confidently capitalist abstraction and commodification of space, but as part of a much more careful rhetorical rhythm of projection and limitation. Reformist schemes such as Dymock's for the English farm or indeed Milton's for education can be regarded similarly as tentative attempts to balance spiritual virtue and intellectual truth against human frailty and acquisitiveness through the mediating discipline, or 'reduction' of good 'design'.

Milton, as Bruce McLeod acknowledges, was wary of and explicitly condemned both commerce and empire (McLeod 1999: 134). He figured the chaos of unfettered individualism and material desire both in Satan and in the great belly-God Comus, and whilst he may have aspired in his later writing to a generous vision of paradise regained within the individual regenerate soul, in the 1640s he still saw the need for a rigorous discipline to mitigate man's inherent corruption (Stavely 1987). In *Reason of Church-Government* (1642) Milton observes that 'the state also of the blessed in Paradise, though never so perfect, is not therefore left without discipline, whose golden survaying reed marks out and measures every quarter and circuit of new Jerusalem' (*Reason of Church Government*, Milton 1953–82, I: 752). In his production of 'designs', whether poetic or social, Milton sought to mediate a relationship between universal and particular that would regulate and redeem the fallen world.

An implicitly or explicitly mathematical and geographic language of design, surveying and reduction was common currency amongst Milton's Puritan peers, for whom it served the same rhetorical function of negotiating between virtue, grace and providence and the chaotic energies of history and commerce. Towards the end of his life, the Puritan George Wither wrote on the Dutch wars of the 1660s, some of whose less glorious episodes were viewed by many of his generation as a marker of the corruption and effeminacy of the Restoration government. Wither's *Three Private Meditations* were written at the conclusion of 'the late Ingagement between the English and the Dutch, in June 1665' (*Three Private Meditations*, Wither 1872–77, IV: 1). Wither gives thanks to God for his 'mercy' to the English in their conflict, and for his 'providential respect' to Wither and his family, imprisoned in Newgate for three years at the Restoration, and expresses a more general appreciation of divine power and justice. To mark this omniscience and omnipotence, Wither figures God as a surveyor: an elevated and disinterested geographer of human quarrels:

> Who lookst down from Heav'n, upon
> All that here on Earth is done,
>   And survey'st her darkest parts,
>   Ev'n the Crannies of mens hearts.
>
> ...
>
> Thou observest what was done,
> Not in *Africa* alone,
> Or both *Indies*; but what was
> Done likewise in ev'ry place:
> Why the *English* and the *Dutch*
> Are divided now so much.
>   (*Three Private Meditations*, Third Hymn, 'A Private Thank-Oblation',
>   Wither 1872–77, IV: 12)

Throughout his long writing career Wither appealed consistently to the figure of surveying to express this kind of impartial judgement on worldly affairs which yet was no retirement from them. *Britain's Remembrancer* (1628) deals with the plague of 1625, drawing morals for the benefits of 'Christian neighbourhood', and taking the opportunity to chastise human improvidence and licence. In canto two Wither figures himself as an observer walking amongst the Whitsuntide festivities of 'Country Townes' – maypoles and hobby horses – and draws upon the figure of surveying to inspire a reforming perspective upon such foolishness:

> For when on Contemplations wings I flye,
> I then o're- looke the highest *Vanity*.
> I see how base those fooleries doe show,
> Which are admired, while I creepe below:
> And by the brightnesse of a two-fold light
> (Reflecting from Gods word to cleare my sight)
> Faiths objects to her eyes, much plainer are,
> Then those which to my outward fight appeare.

My towring *Soule* is winged up, as if
She over-flew the top of *Tenariffe*,
Or some far higher Mountaine; where we may
All actions of this lower World survey.
(Wither 1880: II, pp. 45–6)

There could be no clearer illustration than this of how much Puritanism owes to Plato, despite its rejection of speculative philosophy; oscillating as it does between asceticism and the world. For Wither, surveying represents the capacity of the faithful to see beyond the worldly errors of custom to an ideal, graceful pattern through which the world may be reformed: a perspective to look beyond tradition. Yet this transcendent perspective risks no loss of material detail. It is no Platonic escape from the world, but a reforming, improving geography.

Further into the poem, Wither plays the surveyor again. In canto four, he recalls the arrival of the plague in London. Bemoaning the short memories of those who have returned to their pursuits without giving adequate thanks to God, he seeks to picture the drastic changes which the plague wrought on a busy and pleasure-seeking town. Hyde Park and Marylebone are pictured on the eve of disaster, full of idle maids and gallants; then the London of trade and business:

And, when I did from some high Towre survey
The Rodes, and Paths, which round below me lay,
Observing how each passage thronged was
With men and Cattell, which both wayes did passe;
   How many petty paths, both far and neare,
With rowes of people still supplied were;
What infinite provision still came in,
And what abundance hath exported bin.
(Wither 1880: IV, pp. 242–3)

Wither's elevated viewpoint serves the purpose not of contemplation only, but of a detailed inventory of his subject. As a poetic surveyor of his nation, Wither sees and embraces it in all its local material productivity; its 'infinite provision'; and yet brings all this 'abundance' within a godly reforming frame.

Reduced within their bounds: American improvements

Far from being a powerful means of emptying and ordering the world, cohering neatly with the geometric rationalization of domestic and colonial space, I think the doctrine of improvement was constrained in the seventeenth century by an anxious awareness of the dubious morality of artful individualism. It advanced only by a careful rhetoric of self-discipline and limitation. Early modern writing on America was enduringly obsessed, as Anthony Pagden puts it, with 'the deleterious consequences of over-extension' (Pagden 1995: 161). This enduring obsession can be explained in the early years of settlement in terms of a paranoid garrison mentality which sought security in close confinement, and in terms of a mercantilist pragmatism which sought profits in an efficient oversight of production (Webb 1979). Yet William Bradford feared the scattering of his colony not simply because such a scattering heralded the decay of his hegemony as governor and shareholder in the plantation, but because such unfettered individualism undermined a moral economy defined by limitation.

Describing the process of improvement by which the colony expanded, Bradford writes:

> The people of the plantation began to grow in their outward estats, by rea[son] of the *flowing* of many people into the cuntrie, espetially into the Bay of the Massachusets; by which many were much inriched, and commodities grue plentifull; and yet in other regards this benefite turned to their hurte, and this accession of strength to their weaknes. For now as their stocks increased, and the increse vendible, there was no longer any holding them together, but they must of necessitie goe to their great lots, they could not other wise keep their katle; and having oxen growe, they must have land for plowing and tillage ... By which means they were scattered all over the bay ... and those that had lived so long together in Christian and comfortable fellowship must now part and suffer many divissions.
>
> (Bradford 1946: 293)

Conspicuous here is not simply a false consciousness masking capitalist with moral discipline, but a genuine tension between what Peter Carroll calls 'two logically antithetical versions of the mission to New England' (Carroll 1969: 3). On the one hand the Puritans aspired to create an enclosed and undivided community modelled on the Medieval town; on the other they defined their venture in terms of the ongoing improvement of uncultivated land (ibid.: 3). Throughout the seventeenth century, states Carroll emphatically, neither version of the Puritan mission 'established hegemony over the other. Instead, both versions of the wilderness community – the cohesive and the expansive – flourished simultaneously' (ibid.: 3).

Conditioned by this intense ambivalence, the emphasis we often find in early American writing on spatial discipline is less a stable ideology of authoritarian control and more a tentative rhetoric of persuasion, promoting virtuous settlement in the eyes of an often sceptical domestic audience and negotiating a palatable identity for a commercial yet ascetic people. The disciplinary culture reproduced in their practices and aesthetics helped seventeenth-century Puritans to see beyond the 'declension' traditionally associated with refinement, providing a perspective to look beyond tradition

(Peterson 2001: 336). Exemplary here is an anonymous New England tract titled 'Essay on the ordering of towns' (*c.*1635), which projects a social design highly comparable to Cressey Dymock's near-contemporary model farm.

The essay seeks to establish 'comportable Communion' in the embryonic Puritan community through a plan 'square 6 miles euery waye. The howses orderly placed about the midst, especially the Meetinghouse, the which we will suppose to be the Centor of the wholl Circomference' (Anon. 1943: 182, 181). Geometry here is neither simply a pragmatic mode of laying out the standard town, nor does it simply clear the land for private property. Rather it moralizes the expansive work of settlement through limitation. The essay is certainly preoccupied with a 'speedy Improvement' of the ground incorporated within its design, regarding improvement as a 'princi-pall Condicion of that Grand Couenant assigned' to man by God, without which no man has any 'Theologicall Right vnto any possession ... what pre-tence of Civell Right soeuer he may challenge vnto himselfe' (ibid.: 182). It also regrets the necessity to lay out land in common, rehearsing the conven-tional wisdom of enclosure polemic that 'one acre inclosed, is much more beneficiall then 5 falling to his share in Common' (ibid.: 184). Yet at the same time the author assures his reader that all 'within Compas of the wholl towne' will be 'bownd with the suerest Ligaments'; each man limited to 'his due proportion' (ibid.: 184, 183). This spatial morality was re-invoked where New England towns began to test their bounds.

In 1667 the residents of the southern, Chebaco district of Ipswich town petitioned Massachusetts General Court successfully for parish status, sup-porting their application with a plan (Benes 1981: 57; Fairbanks and Trent 1982: 34) (Figure 4.2). The plan showed the proposed site for a new meeting house at the centre of an octagonal shape. This geometry conveyed a distinctly spatial, and at the same time distinctly moral, rhetoric. It was designed to demonstrate that none of the houses of the new parish would be more than two and a half miles distant from 'comportable Communion', whereas some of them were currently seven and a half miles distant from the Ipswich meeting house.

In the first histories and travelogues of New England the oscillation between expansion and limitation made graphic in the Chebaco plan is played out at length. This oscillation, rather than the confidently expansion-ist trajectory across a blank Cartesian plane, is the true rhythm of American improvement.

Captain Edward Johnson came to America in 1630 with John Winthrop as a joiner. He was granted a large tract of land in Charlestown in 1638 and then helped found the new town of Woburn in 1640. Johnson was made town secretary of Woburn and was frequently employed as a skilled surveyor by Massachusetts General Court. Johnson's *Wonder-Working Providence*, written mostly in 1650 and published in 1654, locates the founding of Woburn and the other New England towns within a framework as much

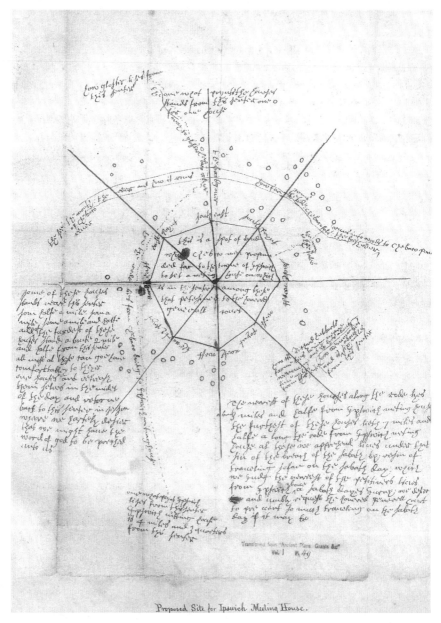

Proposed Site for Ipswich Meeting House.

Figure 4.2 Unknown draftsman: Map of Chebaco, Massachusetts Archives Collection Third Series (v.1 p.35), 1667. Reproduction courtesy of Massachusetts Archives.

defined by providentialist chronology as spatial geography. This limiting framework allows him to map the progressive 'improvement' of the New England wilderness with a delight in its growing prosperity. Johnson's geographic mode slips rhythmically from typology to celebratory chorography, as in this description of Boston:

> But now behold the Admirable Acts of Christ; at this his peoples landing, the hideous Thickets in this place were such, that Wolfes and Beares nurst up their young from the eyes of all beholders, in those very places where the streets are full of Girles and Boys sporting up and downe, with a continuall concourse of people. Good store of Shipping is here yearly built, and some very faire ones ... this Town is the very mart of the Land, French, Portugalls and Dutch come hither for Traffique.
>
> (Edward Johnson 1974: 71)

Johnson's perspective on the Puritan venture can often seem every bit that of the accountant, reducing it as the English surveyor does the manorial estate to 'facts and figures'. But he is consistently careful to bring the material improvements he describes back within the framework of a moral economy. In Chapter XIII he provides a 'short survay' of the cost to date of settlement (Edward Johnson 1974: 56):

> The passage of the persons that peopled New England cost ninety thousand pounds; the Swine, Goates, Sheepe, Neate, and Horse, cost to transport twelve thousand pounds besides the price they cost; getting food for all persons for the time till they could bring the Woods to tillage amounted unto forty-five thousand pounds ... Assuredly here it lies in banke, put out to the greatest advantage that ever any hath beene for many hundred of yeares before.
>
> (Edward Johnson 1974: 54–5)

Johnson's assurance that improvement is a mark of grace does not dissipate the tension that Peter Carroll describes as integral to the Puritan ethos: a tension that marks and defines the seventeenth-century 'discourse of improvement'. Whilst he emphasizes at every stage the virtuous struggles of his brethren – 'the laborious breaking up of bushy ground, with the continued toyl of erecting houses ... in this howling desert' – Johnson is also careful to bring such improvements within disciplinary limits which will absolve his brethren from the charge of greed and individualism:

> Then all of you, who are now, or shall hereafter be shipped for this voyage, minde the worke of Christ ... others eying the best Grasseplatts and best situaion for farmes and large Accomodations, crouding out Gods people from sitting down among you. Wherefore above all beware of covetousnesse ... Lastly, let not such as first set foote on Land

to compose Townes for Habitations, take up large accommodations for sale, to inrich themselves with others goods, who are to follow them, but freely as you have received, so give out to others; for so soone as you shall seeke to ingrosse the Lords wast into your Lands, he will ease you of your burden by making stay of any farther resort unto you, and then be sure you shall have wast Land enough.

(Edward Johnson 1974: 35)

Rather than displacing a traditional Protestant discourse of complaint against enclosure, Josselyn's chorography of Puritan improvement is counter-pointed by it, expressing the same ambivalence and following the same tidal rhythm as Wither's survey of the abundance of London. This ambivalent perspective from within the Puritan colonies is to some degree mirrored by a far more liberal perspective from without, where a disciplinary rhetoric works once again to restrain rather than promote the expansive energies of improvement.

John Josselyn paid two visits to America, in 1638–39 and 1663–71, on both occasions visiting his brother Henry, a commissioner working under the partnership of Sir Ferdinando Gorges and Captain John Mason in Maine. His account of what he saw in *Two Voyages to New-England* (1674) is in many ways a celebration of colonial improvement, expressed in markedly carto-graphic terms. As he describes the New England landscape Josselyn seems to be moving his finger over a map, covering the bones of distance and toponymy with the flesh of successful settlement and prosperity:

From *Connecticut*-River Long-Island stretcheth it self to *Mohegan* one hundred and twenty miles, but it is but narrow and about sixteen miles from the main; the considerablest Town upon it is *Southampton* built on the Southside of the Island towards the Eastern end ... The Island is well stored with sheep and other Cattle, and Corn ...

Cape-Cod was so called at the first by Captain *Gosnold* and his Company ... and afterward was called *Cape-James* by Captain *Smith*.

(Josselyn 1988: 110–11)

But notwithstanding this celebration of English improvement in every form from corn to the geographic measurement and naming of the land-scape, Josselyn leaves room for Indian geography. He divides his account into two parts: a highly specific cartography of English settlement, as exem-plified above; and a far more generalizing, scarcely localized survey of native life. Not only does he keep alive hopes that the Indians will assimilate, following the good examples of English settlers, but he pays some regard to a customary Indian geography preceding assimilation, paying homage to the Restoration Crown's imperial recognition of the Indian tribes as sovereign nations. Indeed Josselyn plays directly to anxieties about the loyalty of the Puritan colonies, suggesting that the Indians of Massachusetts 'may in good time be known for honest Kings men' (Josselyn 1988: 125).

Indian rights serve Josselyn as part of a rhetoric of government through limitation. Josselyn probably received patronage from the Crown in recognition of the role his report played in settling the dispute of Gorges with Massachusetts over their boundary. In his discussion of this dispute, and of the dispute between England and Holland over New Amsterdam/New York, Josselyn constructs the relationship of sovereign crown to colony much as writers from Bradford to Johnson construct the relation of colonial government to settler: as a rhythm of expansion, scattering, transgression; disciplined by constraint, limitation, measurement. In 1664, writes Josselyn of the New Amsterdam dispute, Charles II sent commissioners 'to reduce the Colonies into their bounds, who before had incroached upon one another', and again, when Massachusetts 'incroached upon' Gorges settlements in Maine, 'his Majestie ... sent over his Commissioners to Reduce them within their bounds' (Josselyn 1988: 108, 136). In a description of Boston which sits interestingly alongside Johnson's he echoes this rhythm of expansion and reduction, allowing himself to admire the town's own exemplary balance of prosperity and constraint:

> The houses are for the most part raised on the sea-banks and wharfed out with great industry and cost, many of them standing upon piles, close together on each side the streets as in *London*, and furnished with many fair shops, their materials are Brick, Stone, Lime, handsomely contrived, with three Meeting Houses or Churches, and a Town-House built upon pillars where the Merchants may confer, in the Chambers above they keep their monethly Courts ... The Town is rich and very populous, much frequented by strangers ... On the south there is a small, but pleasant Common where the Gallants a little before sun-set walk with their *Marmalet*-Madams, as we do in Morefields, &c. till the nine a clock Bell rings them home to their respective habitations, where presently the Constables walk their rounds to see good orders kept, and to take up loose people.
>
> (Josselyn 1988: 114)

Those, from Bradford and Winthrop to Johnson and Josselyn, who advanced the claims of a labour theory of economic value and cultural identity in the seventeenth century were typically careful to circumscribe and qualify their claims in much the manner made graphic in the Chebaco plan; to define the morality of improvement through limitation. If mathematics was entangled with a discourse of improvement in the seventeenth century it was not generally as a tool of dominion and erasure but as part of a rhetoric of constraint. The labour theory, as Locke himself conceived it, limited man's rights to property to the amount he could improve, leaving no spoilage, and leaving enough good land for the industrious use of others (Arneil 1996: 102–3). Even where he describes the labour theory in its purest form, Locke, like Milton, uses a rhetoric full of notably mathematical circumscription.

'The measure of Property,' he writes, 'Nature has well set, by the Extent of Mens *labour* ... This *measure* did confine every Man's *Possession*, to a very moderate Proportion' (Locke 1967: II. 36, p. 292). Moreover, Locke's theory of property and civil society did not leave the measure of property to Nature alone.

Locke argued that whilst labour 'begins' property in the natural state, it is the responsibility of civil governments to regulate and circumscribe such property: to set its bounds (Arneil 1999: 157–8). Locke's work for his patron Shaftesbury in drafting the Fundamental Constitutions of Carolina secured the rights of the individual colonist, but was still more concerned with 'the orderly management and division of the land by the government' (ibid.: 128). In the administration of the colony Locke acted to limit the capacity of individuals to claim land through labour, insisting in 1672 on 'reducing' settlement to the 'bounds and limits' established in surveys commissioned by the Lords Proprietors (quoted in Arneil 1999: 129–30). In setting such paternalist limits to improvement he sought avowedly, as earlier colonial governments had, to keep the peace with and to protect the claims of those potential co-improvers the Indians, and thereby to keep the peace with the English Crown, whose role in disciplining its improving colonists his model of settlement and appropriation accommodated (Arneil 1999: 161–2).

If mathematics defined improvement in a Lockean age, it also, perhaps more importantly, constrained it. And as it did so, it worked not against but in concert with a traditional patriarchal ethos pretending to the stewardship of customary rights. This negotiation between stewardship and improvement is the subject of the next chapter.

5   Ambivalent geographies

With John Josselyn's highly cartographic account of New England we move from general rhetorics of mathematics and social discipline to discourses directly engaged in the business of geography. Having foregrounded the ambivalence of a Protestant culture of discipline I now want to examine the role of geography in negotiating this ambivalence, and in building bridges between the truths and values of a post-feudal moral economy and the truths and values of individualism and reform.

Faultlines

Not all historians agree with the conception of estate surveying as an inherently improving genre, or with the notion of a common Protestant expansionism uniting estate maps and the geographies of the imperialist nation state. Bernhard Klein detects a faultline in surveying between a geometric aesthetic of circulation – 'the idiom of the market' – and the particularities of authority, custom and place which define 'the integrity of the estate' (Klein 2001: 54–6). Richard Helgerson suggests a geographic rift between Crown and country. For Helgerson early seventeenth-century estate surveys, like country-house poems of the period, carried a 'potent ideological charge' directly contrary to the antiquarian chorographies of the post-Saxton era (Helgerson 1993: 69). They supported the 'neo-feudal' Caroline and Jacobean conception of the patriarchal manorial economy as first principle of the order of the state (ibid.: 70). Where chorography allowed the land itself to 'speak', to a relatively democratic community of educated consumers, early seventeenth-century surveys and country-house poems subsumed the land and its inhabitants within the political structures of lord, and ultimately King. The Elizabethan shifting of crown investment from national topography to the surveying of its own estates, suggests Helgerson, represented an attempt to strike a blow on its own behalf in 'an implicit contest for state power' (ibid.: 69).

Caught in the shifting, unstable politics of early seventeenth-century geography, as characterized by Richard Helgerson, was the surveyor John Norden. Remembered principally for his topographical work, Norden was authorized by Queen Elizabeth to travel through England and Wales 'to make more perfect descriptions, charts, and maps' (quoted in Gerish 1903: ii). He was never salaried by Elizabeth, but enjoyed the initial support of a range of patrons, and Crown interest was 'not far behind', indicated in the liberal passes he secured (Barber 1992: 64).

In 1593 Norden published his first fulfilment of this commission: *Speculum Britanniae: The First Parte. An Historical and Chorographicall Discription of Middlesex*. His description of Hertfordshire followed in 1598, preceded by a *Preparative*, published in 1596, which sought to defend his methods in producing these topographies (Norden 1596). Ultimately, however, the Crown withdrew its support for Norden's topographic project, refusing to renew his pass. Instead, it supplied him with the best rewarded and most productive employment of his career, making him in 1600 surveyor of Crown woods and forests in the southern counties, and then in 1605 surveyor to the Duchy of Cornwall.

In these capacities Norden surveyed according to his own estimate 176 manors, many more than once, and earned an average of around £200 a year, contrary to his own frequent complaints of poverty (Kitchen 2004). He supplemented this income, when surveying work was slack, as a poet and as a promoter of his own art of surveying. The work of Norden's that has attracted the most attention from cultural historians in recent years is not a work of 'real' topography but a fictional construction of the surveyor's persona, designed to instruct in and defend the art. In this text we catch a glimpse of the truly equivocal nature of early modern geography, defined by the competing values of traditional stewardship and improvement.

Although *The Surveyor's Dialogue* (1607) contains guidance in some advanced mathematical techniques and instructs in the making of a map, Norden's surveyor is still very much the older type. To recall Edward Worsop's delineation of the various aspects of surveying, Norden is concerned more with an understanding of the 'diversities' and particulars of law and custom than with universal science; more with empirical judgements about the singular 'nature'; the propriety of 'every kinde of ground' than with the geometric constants of 'proportion and symmetrie'; more with the words of records, books, interviews and courts than with the new surveyor's cartographic mathematics. Asked whether he will measure with a long-range instrument, or with chains, Norden's surveyor replies that he will use both. For a man may well, 'having the true use of any topographicall instrument by rules geometricall, describe a Mannor in a kind of forme, without line or chaine, or other measure. But if he will say he doth, or that he can truly delineate a Mannor with all the members, as every street, high-way, lane, river, hedge, ditch, close, and field ... I will then say he is *rara avis*' (Norden 1607: 125). Only through direct, empirical contact with the particularity of his subject, insists Norden, may the surveyor 'produce an exact discoverie and performance of the worke he undertaketh' (ibid.: sig.B2v). And this close engagement includes traditional processes of verbal inquisition. 'These are the two pillers,' states Norden, 'upon which a Surveyor must of force build his worke, information and record' (ibid.: 23).

The Surveyor's Dialogue reminds us that Norden's era was witnessing only a gradual and partial separation of the work of surveying from that of day-to-day manorial stewardship, and places this work within the context of distinctly pre-capitalist moral values regarding land use. Norden's introductory dedication to Robert Cecil, Secretary of State under Elizabeth, and James I's Lord Treasurer from 1608, depicts the surveyor maintaining and defending the customary social order connecting King to commoner through the land. This construction of the surveyor's role is clearly designed to appeal to Cecil's own conservative vision of a fundamentally agrarian order.[1] 'There is none,' warns Norden, 'but well considereth, that how great or powerful soever he be in land revenues, it is brought in unto him by the labours of inferiour tenants: yea the king consisteth by the field that is tilled' (Norden 1607: sig.B2r). 'Dominion and Lordship,' he reminds his reader, 'principally grow ... by Honors, Mannors, Lands and Tenants' (ibid.: sig.A3v). The surveyor can and should enhance the landlord's revenues, but only so that such revenues may be 'the meanes to enable the Honorable, to shelter the virtuous distressed, and to cherish such, as by desert may challenge regard' (ibid.: sig.A3v). Surveyors should help Lords to reap a 'reasonable improvement', whilst dealing justly with those who come 'within the compasse of their reuenues' (ibid.: n.p.). Norden's 'compassing in' of profit does not make room for enclosure, which he deems 'the bane of a common wealth' (ibid.: 218). And where Norden writes as a surveyor for the Crown it is not the wasted ground of unexploited commons he complains of, but the waste of customs that bind his post-feudal world together. 'It is an offence,' he laments, in a discursive part of his manuscript *Survey of Prince Charles's Manors*, 'to breake reasonable and lawfull customs, to the prejudice of poore tenauntes' (Norden 1617: sig.15r).[2]

Richard Helgerson's account of Norden as chorographer and surveyor presents him as working successively against and with the grain of two distinct genres of representation to accommodate the conservative social vision of the Crown (Helgerson 1993: 72). Other commentators, however, have perceived an acute tension within Norden's conception of surveying. Both McRae and Klein emphasize the ambivalent nature of *The Surveyor's Dialogue*, describing it as 'tangled in a web of anxieties and uncertainties', and 'fraught with ideological tensions', and tracing in it a faultline between the 'idiom of value and profit' and a 'residual feudal terminology' (McRae 1996: 178, 179; Klein 2001: 57). McRae characterizes Norden's text as 'one of the most extensive attempts to negotiate a position between the old social morality and the new standards of knowledge and reason' (McRae 1996: 177).

Indeed just as he equivocates over the universal overview of geometry, Norden also equivocates between conservation and change. He concedes that some wastage of custom may bring about positive effects. The 'consuming' of woods in the wealds, he reflects, is after all 'no such great prejudice to the weale publike', since the 'cleansing' of such grounds benefits the country in two ways (Norden 1607: 215). First, 'where woods did grow in superfluous abundance, there was lacke of pasture for kine, and of arable land for corne' (ibid.: 215). Second, Norden adds, 'people bred amongst woods, are naturally more stubborne, and uncivil, then in the Champion Countries', and have more need 'to seeke the meanes of reformation' (ibid.: 216). And if Norden's definition of the values of surveying was fraught with ideological tension, so, *pace* Helgerson, was his actual surveying work for the Crown.

From 1604 there were 125 men, including prominent publicists of the surveyor's art such as John Norden and Aaron Rathborne, employed in the 'Great Survey' of Crown lands, designed to enhance income by improving records (McRae 1996: 174–5). McRae regards the debate which produced and surrounded this survey, as usual a struggle between advocates of custom and reform, as a crucial element in the formation of 'the status and discourse of the modern surveyor' (ibid.: 175). In the course of this debate, and in the course of the surveys themselves, the Crown modified somewhat its position as arbitrator and protector of the rights of tenants against unscrupulous landlords. Whilst it continued until the end of Charles I's personal rule to arbitrate disputes with considerable sympathy for the rights of tenants, the Crown's actions in its own property interests came increasingly, from the reign of Elizabeth I onwards, to resemble those of the speculative, improving landlord (Crossley and Willan 1941: xiv). In this context it is hard to share Richard Helgerson's interpretation of Norden's surveying work as in any unproblematic way integral to a conservative Jacobean socio-economic vision. As David Buisseret comments, 'when John Norden drew the maps of James I's Windsor estate ... he was acting for James not as King but as a great landowner' (Buisseret 1996a: 1).

Norden's disavowal of reform, and indeed his disavowal of mathematics, was part of a fraught negotiation between old and new truths and values in which it is impossible and misleading to disentangle the characteristic politics and aesthetics of individual artists, sponsors or generic modes. The tension Bernhard Klein notices between an improving, economically expansionist mathematics and an aesthetic and ideology of locality, custom and tradition is a tension which parts much recent scholarship on early modern geography, producing conflicting, often polarized readings of its characteristic epistemology and politics. Yet I think this tension was itself a highly significant, even a formative element in seventeenth-century English geography and in the political imaginary which it informed. If English geography was at every level a 'dialectical' discipline, as Helgerson terms it, this dialectic involved an anxious equivocation between particular, local custom and reform.[3] English geography was characteristically neither conservative nor 'whiggish', to use Helgerson's consciously anachronistic term. It was ambivalent: overdetermined by shifting, overlapping determinations of what was true and what was virtuous in the use and representation of space.

That surveying remained a heterogeneous, ambivalent art throughout the seventeenth century has been demonstrated in fine-grained archival research. Evolving from the military use of scale plans, and the administrative use of large-scale cartographies such as Saxton's, the genre of estate mapping emerged from the 1570s onward fully fledged, some of its most sophisticated examples appearing almost immediately (P.D.A. Harvey 1996: 27–9). Yet this early and immediate transformation of the work of land management and representation was by no means comprehensive: it simply offered new possibilities to surveyors and their employers. The results of the surveyors' work might be compiled within a map, replacing a traditional written survey; both map and survey might be produced, with various relations possible between them, or the written survey might still stand alone (ibid.: 38). The majority of seventeenth-century landowners, suggests P.D.A. Harvey, probably did not commission maps at all (ibid.: 41). Those maps that were produced were generally based on elementary instruments and techniques, rather than the sophisticated theories and practices promoted in contemporary manuals, and remained 'only one tool of estate management' in a period and area in which 'terriers, valuations, and written surveys continued to be important and often remained more numerous than estate maps' (Bendall 1996: 70).

Early instances of the comfortable relationship between the new cartography and the traditional arts of manorial stewardship and written survey can be seen in the work of a family of late sixteenth- and early seventeenth-century surveyors: the Walkers of Hanningfield in Essex. Twenty-two maps and two written surveys surviving largely in the Essex County Records Office can be attributed to one or both of the two John Walkers, along with a few others by close associates. In the Walker maps, produced in the relatively liberal socio-economic environment of East Anglia, cartography sits comfortably within a traditional surveying ethos.

Whilst asserting the truth and perfection of their mathematics, the Walker surveys are also scrupulous in their stewardship of the customary integrity of the manorial estate, compassing the myriad details of their legal and physical identities within the 'just proportion' of a mathematical framework.[4] John Walker senior accompanies 'A Trew Plat of the Man[n]or And towne of Chellmisforde 1591' with a written survey book, described by him as follows:

The Manor of Chelmersforde. A booke of the Survey and admeasurem[en]t of the saide Mano[u]r, demeasnes & s[e]rvic[es], lib[er]ties franchesies & other hereditanm[en]t{es} of the saide Mano[u]r by exact views of the same, vppon the searche of the Courte rolls, rentals, and other Material escriptes of the said Mano{u}r, at the Courte Leete, & Court Baron there holden for the righte worshipfull S[i]r Thomas Mildmaye Knighte, on Tuesdaye the twentieth daye of June in the three and thirtith yere of the reigne of o[u]r sou[er]eigne Lady

Quene Elizabeth before Edward Moryson Esquire S[ur]veyo[u]r John Lathum gent Steward, & John Walker Measurer, Rob[er]te Wood and other ten*an*tes and suito[u]*rs* there.

(Edwards and Newton 1984: 45)

John Walker junior prefaces his 1614 survey of Foxcott Manor in Andover, Hampshire with the following confident assertion:

All which sayde Mannor landes, arable, pasture, meadowes and woodes with the Sheepdowne aswell the copie lands as demeines howsoever they lye dispersed either about fosscutt or Hatherden belonginge to the sayde Mannor in common or otherwise are sett forth in true and just proportion with euerie, Gate, Style Pathe, Ponde, Brooke, Bridge, Highwayes and Driftwayes leadinge into through or by any parte peece or percell of the sayde landes.

(Edwards and Newton 1984: 65)[5]

Contemporary with John Norden's apparently ambivalent definition of the surveyor's role, the dialectical nature of the Walker surveys, like that of Norden's manual, is by no means unique to the very early seventeenth century. These surveys embody a flexibility between custom and improvement; words and mathematics inherent to seventeenth-century geography. P.D.A. Harvey provides instances of the same 'conservative' uses of combined maps and written surveys from the late seventeenth century, as well as instances of written surveys used alone throughout the period, recording enclosures as well as persisting Medieval tenures (P.D.A. Harvey 1996: 38–9).

To some extent the increasing divorce discerned by historians between cartography and the writing practices and legal context of traditional stewardship may well be an illusion produced by history itself. 'Because they are picturesque or quaint,' writes Harvey, 'estate maps are often divorced from the context of the other records, even from the written surveys that they were meant to accompany' (P.D.A. Harvey 1996: 58). If these maps marginalize or erase the customary social landscape, their original users may well have been able to find it elsewhere.

Limiting the labour theory of value

I think an understanding of seventeenth-century geography as the Baconian disciplining and evacuation of post-feudal space seriously underestimates what cartographic historians have shown to be a slow and uneven negotiation between traditional and newer practices and values. To understand early modern geographies less reductively we must avoid at all costs detaching them from their contexts, and set limits to our presumptions about the inherent politics of various modes of using and representing space.[6] The first thing we need to set limits to, to understand the ambivalence of early modern geography, is a Lockean labour theory of value.

Over the course of the seventeenth century, writes Bruce McLeod, 'a labor theory of value was inexorably becoming the justification for expropriating land inhabited by others ... Thus wilderness and commons alike were essentially up for grabs since both lacked the organizing principle of private property' (McLeod 1999: 92). This is simply not the case. Despite his avowed intent to demonstrate the absolute right of property gained through improvement, Locke himself backtracks for a moment at one place in his chapter 'Of property'. 'Tis true,' he writes:

> in *Land* that is *common* in *England*, or any other Country, where there is plenty of People under Government, who have Money and Commerce, no one can inclose or appropriate any part, without the consent of all his Fellow-Commoners: Because this is left Common by Compact, *i.e.* by the Law of the Land, which is not to be violated. And though it be Common, in respect of some Men, it is not so to all Mankind; but is the joint property of this County, or this Parish.
>
> (Locke 1967: II. 35, p. 292)

This qualification seems to topple Locke's entire argument if we understand it, as commentators generally do, as primarily designed to ground English civil identity on exclusive private property, regulated *post facto* by common consent, and to discount all other customary arrangements, from post-feudal copyholds to the hereditary monarchy. Locke reminds us that most enclosure in the seventeenth century was not prosecuted according to some putative labour-theory right of improvement, whereby inefficiently yielding common land was forfeited by its customary users in favour of an industrious and more artful private farmer, but redistributed consensually amongst its present users, however coercive the process of obtaining such consent.[7] Locke found it no more easy than his contemporaries utterly to discount the customary rights of the English commoner. He would certainly have been dismayed to think that his thesis left the English commons simply 'up for grabs'.

The notion of geography as the disciplinary tool of the improver might seem to make more sense for America. Where Locke uses the term 'wast' in the *Two Treatises*, concludes Arneil, he does not intend us to understand 'open field tillage in England' as his editor Peter Laslett translates it (quoted in Arneil 1996: 7). It is *American* waste that Locke defines as land 'left to Nature' (Locke 1967: II. 37, p. 294); indeed, so definitive of the virgin state is this land that 'in the beginning,' he says, 'all the World was *America*'. And the right of the American colonizer to appropriate this waste might well seem absolute because those competing rights conceded to the English commoner are missing. The inhabitant of America, notes Locke, whilst he is certainly, lacking private property, still a 'commoner', does not only lack the rights enshrined in English law; he has not even 'joyned with the rest of Mankind, in the consent of the Use of their common Money' (ibid.: II. 45, p. 299). Yet I think we should hesitate before we join with McLeod and others in marking the confident and speedy evacuation of native America through a geography recognizing only Lockean property and improvement.

Medieval deeds and impotent geometries

Ireland and America both furnish substantial colonial examples of the reduction of space to disciplined, manipulable form that seventeenth-century surveying manuals seem to describe. There is no doubt, for instance, that the extensive surveying and cartographic operations overseen in the middle and late seventeenth century by William Petty in Ireland and Thomas Holme in Pennsylvania are exemplary of such colonial reductions.[8] They confirm Harley's thesis of a 'dehumanised geometrical space ... whose places could be controlled by coordinates of latitude and longitude'. But this aspect of seventeenth-century colonial geography is by no means as typical as it is often made to seem. As well as Cartesian abstraction and Lockean improvement, seventeenth-century geographies of America, like their English counterparts, routinely accommodated more traditional knowledges and values.

Detailed archival studies of local American practices in the use and representation of land have suggested a profound conservatism in many areas. These studies have drawn attention to documents which belong more to the Medieval open field system than to a modernity of discipline and improvement.

The documents which had most bearing on land ownership and use in seventeenth-century America were not maps at all, but old-fashioned verbal deeds. Margaret Wickens Pearce has argued that the deeds recording trans-actions between colonists and Indians, hundreds of which are still preserved in old colonial archives, represent a buried tradition in early American geo-graphy (Pearce 1988). Pearce blames the burial of this tradition on histor-ians who privilege not what was most important in the seventeenth-century negotiation of relationships to the land, but what best fits a modern, Enlightenment idea of geographic form and function. In most instances, the negotiations of which these deeds formed a part followed procedures and produced documents best understood not within the Enlightenment tradi-tion of mathematical surveying and cartography, but within the older, Medieval tradition of manorial investigation and report.

G. Malcolm Lewis's introduction to Pearce's essay treats her discoveries as highly counter-intuitive (Lewis 1998). If this is so it is because they reveal two things. They reveal a conservatism which sits uncomfortably with our sense of the Enlightened nature of seventeenth-century geography. They also reveal a concomitant preparedness to negotiate consensual uses of American land which sits uncomfortably with our sense of geographic ideology. These documents, in short, are a far cry from the disciplined spaces laid out in Petty's and Holme's Irish and Pennsylvanian geographies.

Even where they did produce graphic maps, New England settlers rarely defined their properties with the mathematical precision we tend to associate with an emergent agrarian capitalism. 'Beyond the usual repertory of printed maps associated with well-known events,' writes Peter Benes, lies 'a body of peripherally known vernacular or "folk" documents that were the product of everyday people who lived below the level of historical scrutiny' (Benes 1981: xv). Much like P.D.A. Harvey and Margaret Wickens Pearce, Benes suggests that our impression of the nature and role of early maps has been distorted by biases in-built to processes of historical survival and preservation. These leave us focusing on 'special exceptions, often decorative or genealogical pieces preserved and handed down by successive generations' (ibid.: xvi). More representative in terms of actual use are simple, barely geo-metrical documents used by the primitive surveyors of seventeenth-century America in the constitution and administration of early settlement.

Although most local governments elected surveyors to 'lay out' individual property bounds and 'return' verbal descriptions of these for the town records, the relatively small holdings of New England, for instance, meant full-blown estate maps weren't necessary for strictly administrative purposes (Candee 1982: 9; Buisseret 1996b: 96). Those maps that were produced were crude, working documents typified in a 1656 town plan of Chelmsford, Massachusetts for which Edward Johnson was partly responsible (Allen 1982: 6) (Figure 5.1). The Chelmsford plan, the earliest surviving for any New England community, sketches in the town between the Merrimack and Concord rivers and next to the planting land of the Indian John Sagamore.[9] David Grayson Allen finds its imprecision typical of a people 'still used to English practice, which relied upon the occasional perambulation of the land and even more on the collective memory of borders and ownership privi-leges', assigned according to the traditional system of verbally naming metes and bounds.[10] 'Even though Massachusetts General Court passed a land recording act in 1634,' he notes, 'it made only modest demands on the towns and placed no requirement on them for an accurate mathematical survey', asking specifically that grants should be 'fairly written in words att length, & not in figures' (quoted in Allen 1982: 6). Rhode Island, less preoc-cupied with 'territorial control and expansion' than its neighbour, passed no mapping laws at all (Benes 1981: 36).

By the latter decades of the seventeenth century a multitude of legal dis-putes had developed in the American colonies over casually determined boundaries.[11] Forced to settle these disputes according to the 'ambiguous or incorrect wording' of lot layers whose mathematical training was often minimal, colonial authorities came at last to call upon the services of experi-enced professional surveyors; to insist that lots be laid out by such skilled persons, approved by themselves, and to 'adopt surveying techniques developed in England a century earlier' (Allen 1982: 8, 34–5).[12] Ultimately,

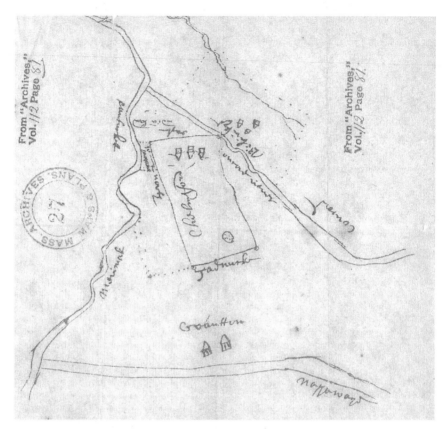

Figure 5.1 John Sherman, Simon Willard and Edward Johnson: Plan of Chelmsford,
Massachusetts Archives Collection (v. 112 p. 81), 1656. Reproduction
courtesy of Massachusetts Archives.

surveying emerged in America not as the foundation for the universal, ideological production of a disciplined and commodified space, but as a simple
practical tool whose use, authority and meaning were highly contingent on
local circumstances. As in England, basic surveys were used in court disputes as often to defend conservative practices of land tenure as they were to
inscribe individual private property, and where they did mark private property they did not necessarily carry the day. Sometimes maps were used to
defend conservative, customary practices against the encroachments of
improvers; sometimes the bounds marked accurately on the surveyor's map
gave way before these same 'Antient' customs (Benes 1981: Figure 83,
pp. 78–9; Allen 1982: 42).

No more conclusively did mathematics mark exclusive property at the
grander level of inter-colonial and inter-state dispute, where bounds were in
some instances hotly contested throughout the seventeenth century and

beyond, despite successive inconclusive surveys. Exemplary here is the case of Connecticut, bounded originally by the Pacific Ocean to the west; Massachusetts to the north, Rhode Island to the east and Long Island Sound to the south. In the words of Clarence Winthrop Bowen, author of a nineteenth-century monograph on the subject: 'the boundary lines have been in perpetual motion since the founding of the Colony' (Bowen 1882: 9). Bowen illustrates this overview with a wonderful quotation from Rufus Choale, who appeared before the legislature of Massachusetts as 'counsel for the remonstrants against the boundary between Rhode Island and Massachusetts':

> Why, gentlemen, the Commissioners might as well have decided that the line between the States was bounded on the north by a bramblebush, on the south by a blue jay, on the west by a hive of bees in swarming-time, and on the east by five hundred foxes with fire-brands tied to their tales.
>
> (quoted in Bowen 1882: 9)

Choale's ridicule is a splendid reminder of the hubris of the colonial patent, which carves up a portion of Euclidean space sprinkled sparsely with an unstable mixture of native and colonial toponymy, and hopes for its declarations to take material effect. In short, the geometry of imperial sovereignty was often comically ineffectual.

Between empire and dominion: geography as public rhetoric

From local property disputes to the rivalry of nations it is clear that conflict and political tumult generated maps. What is not clear is that maps solved conflicts, or were any more than what one historian has called 'gestures in the parry and thrust of political contest' (Gronim 2001: 402). We are wrong to assume that geometry was associated widely in the seventeenth century with the power to reduce domestic or colonial space to empty, manageable form. Instead, I think geometry was part of a much more tentative rhetoric of negotiation, mediating the shifting truths and values of an acutely uncertain age. If geometry was of very little practical use in seventeenth-century America, its significance was greater in this cultural work.

Far from generating the disciplined geometric space of Baconian science and Lockean property, even the best-known, most public geographies produced to promote American colonization were characteristically tentative, negotiating between traditional and newer, proto-Lockean notions of customary and natural rights, and responding to the shifting experiences and priorities of each stage of settlement. Far from erasing the native geography of America they often promoted an idealized partnership of trade and mutual improvement between English settlers and natives whose customs of land use and tenure were presented as far from alien to their own.

Anthony Pagden and David Armitage have demonstrated in recent books the complexity of the thinking in which the earliest European promoters and prosecutors of settlement in America were embroiled. The Spanish worried over whether the *'imperium'*, or imperial authority granted them by papal donation over the peoples of vast swathes of America, gave them *'dominium,'* or 'property rights and sovereignty' (Pagden 1995: 75). The English, in turn, had to produce justifications for the appropriation of American land which would 'beat' both native *dominium* and Spanish *imperium* (Armitage 2000: 92). Most justifications had their drawbacks on one or the other score. Pagden suggests that the first conception of English *imperium* in America founded it on conquest (Pagden 1995: 64). However, he points out that this provided a very insecure grounding for *dominium* when the English widely considered their own rights to own and use their land to have survived the Norman conquest (ibid.: 77). Another avenue for claiming *dominium* lay in the Calvinist argument that it derived from God's grace, and that therefore neither pagan nor Catholic possessed such *dominium* (ibid.: 74). However, both Pagden and Armitage suggest that this argument was rarely used in practice by English writers, who preferred – in Pagden's words – to depict purely religious justification as a 'Catholic aberration' (Pagden 1995: 75; Armitage 2000: 90–1). A final resort lay in the claim of English and other European colonizers to appropriation only of empty, or at least uncultivated, land. This claim drew on both biblical doctrine exhorting man to 'subdue' the earth and the Roman Law tradition which defined *res nullius* (empty things) as common property and a legitimate object of individual appropriation (Pagden 1995: 76). Pagden suggests the ubiquity of the *res nullius* argument in English colonial writing 'from the 1620s on' (ibid.: 77). Armitage points out, however, that whilst it provided defensible grounds for *dominium*, in competition with the native population, *res nullius* was just as available to any other European monarchy as to the English, and thus no rationale for exclusive *imperium* (Armitage 2000: 94–5). Empire required a settled subject population to subject themselves to one, rather than another, superior power. Armitage draws attention to a debate mounted in 1607–08 by the Council of the Virginia Company which weighed up the possibilities for arguing English *dominium* and *imperium* simultaneously in a definitive document, and decided ultimately that such explicit consistency could not be achieved (Armitage 2000: 92–4). Instead, as Armitage puts it, the task of justifying the English actions in Virginia 'would be left to an eclectic cast of promoters, preachers and propagandists' (ibid.: 94).

Conditioned by the uncertainties Armitage describes, the first public geographies of colonial America might well be compared with the Crown surveys with which they were roughly contemporary. These geographies were undoubtedly driven by a Baconian impulse to generate knowledge and thereby profitable improvement through disciplined investigation and dominion over American soil. At the same time they recognized, to recall Norden's words, that imperial 'Lordship' grew by 'Honors, Mannors, Lands and Tenants', and therefore constituted English empire through their recognition of native civil polities. Their geometries were characteristically geometries of empire, rather than geometries of dominion rooted in improvement, and they were often designed to frame and accommodate the native customs and societies of America.

An equivocation between improving dominion and inclusive empire is conspicuous in the earliest rhetorics of English settlement, and it is an equivocation often belied in history and criticism. In 1584 Richard Hakluyt the younger composed and circulated privately his now famous 'Discourse of Western Planting' (Hakluyt and Hakluyt 1935, II). Designed to support Sir Walter Raleigh's attempt to settle a colony in Virginia, Hakluyt's 'Discourse' gives improvement a remarkable prominence in an era still acutely chary about capitalist innovations in the domestic economy. Hakluyt characterizes the America of future English 'planting' as a 'waste firme' (ibid.: 314). This 'waste firm' represents unrealized potential: a blank accounting sheet and cartographic template available for artful filling in. And art is what England currently abounds in, to the point of surplus: 'nowe there are of every arte and science so many, that they can hardly lyve by one another, nay rather they are readie to eate upp one another' (ibid.: 234). The remedy to this inconvenience, a remedy which anticipates the utopian social solutions of the Hartlib era, is to set this surplus skill to work. This done, writes the mathematician Thomas Harriot, sent by Raleigh as surveyor and historian to the first English colony, Virginia promises:

> commodities there already found or to be raised, which will not onely serve the ordinary turnes of you which are and shall bee the planters and inhabitants, but such an overplus sufficiently to bee yeelded or by men of skill to bee provided as by way of traffick and exchaunge with our owne nation of England, will enrich your selves the providers.
>
> (Harriot 1972: 6)[13]

Since Stephen Greenblatt's influential essay 'Invisible bullets' (1981), a series of readings of Harriot's *A Briefe and True Report of the New Found Land of Virginia* (1588) has viewed it in Foucauldian, New Historicist terms as an exemplary exercise in the disciplining of difference; an early dawning of an Enlightenment age of power/knowledge (Greenblatt 1981; see Mackenthun 1997 and McLeod 1999).[14] Thomas Harriot is certainly well known as a scientist so seduced by the all-encompassing potency of mathematics that he sought to reduce even the most humdrum of English domestic problems to its form (Clucas 2000: 111–12). But he was also the man who spent hours with two Indians brought back to England in a bid to make himself conversant with the detail of their language, and thereby the detail of their customs, their manners and their home (Quinn 2000: 14). Likewise, whilst he certainly claims that many American commodities are little laboured over and little used, Hakluyt too acknowledges the human presence in Virginia that must be framed along with its commodities. He is confident, at this early stage, that Virginia's natives can be 'broughte to a civill governemente', and thus make good subjects of the English Crown (Hakluyt and Hakluyt 1935, II: 253). Both Harriot and Hakluyt are concerned to accommodate and 'steward' Indian difference, rather than simply disciplining and erasing it.

The first English ventures in America were driven by mercantile goals rooted in the experiences of the Levantine and Muscovy trade (Nash 1986: 40). The propagandists and adventurers supporting Raleigh's efforts at Roanoke were inspired by the hunt for a northwest passage to the East Indies, and by the prospect of mutual trade with the Indians. They projected, in Hakluyt's words, 'the vent of the masse of our clothes and other commodities, and ... receaving backe of the needful commodities that wee nowe receave from all other places of the worlde' (Hakluyt and Hakluyt 1935: 275). From Raleigh's privately funded enterprise at Roanoke to Christopher Newport's partly state-supported colony at Jamestown, objectives shifted from the establishment of a trading post to permanent settlement (Nash 1986: 43). Yet the best known geographies of early Jamestown do not suggest a consolidated ideology of spatial discipline and Lockean improvement. They still express a geographic ethos persistently negotiating between empire and dominion; traditional Lordship and Lockean improvement.

In 1612 the maverick adventurer John Smith published *A Map of Virginia*, a text that juxtaposes one of the most famous seventeenth-century maps of America with a narrative 'description' of the region incorporating the infant English colony of Jamestown (Figure 5.2).[15] Smith's map, whose giant compasses noisily draw attention to its mathematical frame of reference, might be viewed in the Baconian terms beloved of New Historicism as disciplining the terrain it represents, privileging the abstract rationale of geometry over details of past and present occupation and use and thereby simplifying the act of appropriation. Yet the map also acknowledges, even insists upon, the Indian presence in Virginia. Aesthetically, the Indians pictured might be dismissed as iconologically marginal. They are on the map rather than in it. However, the map is also full of Indian place names.[16] And at its borders John Smith even acknowledges his dependence on local help and narrative testimony in the making of it. His key reads: 'to the crosses hath been discovered; what beyond is by relation'. The narrative 'description' which accompanies Smith's map acknowledges specifically that the help Smith had came from Indians ('A Map of Virginia', in Smith 1986, I: 151).[17]

Smith's verbal geography of Virginia vacillates, as his cartography might be said to do, between generalizing declarations of the land's emptiness and availability for colonial exploitation, and detailed observations of its current, native use. Smith describes terrain 'all overgrowne with trees and weedes being a plaine wildernes as God first made it', and comments that 'heaven and earth never agreed better to frame a place for mans habitation being of our constitutions, were it fully manured and inhabited by industrious

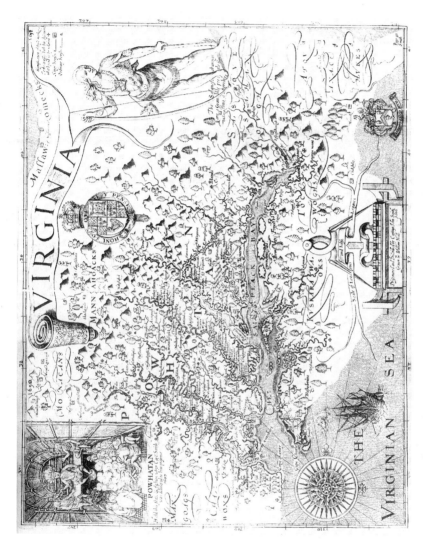

Figure 5.2 John Smith, *Virginia, discovered and Discribed by Captain John Smith, Graven by William Hole* (Oxford, 1612 [1624, 10th State]). Reproduction courtesy of the British Library (shelfmark: G.7120).

people' (Smith 1986, I: 144, 145). Lacking the significant details of English cultivation, the 'plainness' of this well-framed space seems indeed to suggest the inviting emptiness of cartographic geometry. Yet Smith also maps what must seem to the modern reader like the perfectly industrious practices of Indian agriculture, in passages such as the following:

> The greatest labour they take, is in planting their corne, for the country naturally is overgrowne with wood. To prepare the ground, they bruise the barke of the trees neare the root, then do they scortch the roots with fire that they grow no more. The next yeare with a crooked peece of wood, they beat up the woodes by the rootes, and in that moulds they plant their corne. Their manner is this ... Their women and children do continually keepe it with weeding, and when it is growne midle high, they hill it about like a hop-yard. They plant also pease...
>
> (Smith 1986, I: 157)

Moreover these details of native use bring with them, as the Indian contributions to Smith's cartography suggest they should, details of native knowledge: another kind of intimacy with the land. While they have no writing, records Smith, the Indians enjoy their rights 'by customes', and 'all knowe their severall landes, and habitations, and limits' (Smith 1986, I: 174). Far from being empty, Smith's Virginia would appear to be worked, known and owned.

Joyce Chaplin has identified John Smith's sympathetic attitude towards Indian customs with an early stage of settlement, when mere survival necessitated the 'hybridising' of English and native cultures (Chaplin 2001). But beyond the imperatives of survival and the improvement of the settlements, the desire to hybridize European and native cultures was also about the improvement of the Indian. In Smith's writing, as Nicholas Canny observes, the missionary objective is primary and 'sincere', and to be achieved not at the expense of English appropriation and improvement, but precisely through 'the development of a coherent English society in close contact with the natives' (Canny 1988: 215). His incorporation of Indian geography in his verbal and cartographic mapping of Virginia and New England is less a case of disciplined erasure, and more of disciplined intermixture: a framing or 'compassing' of Indian geography analogous to John Norden's 'compassing' of 'Honors, Mannors, Lands and Tenants'. Until the first serious revolt against English settlement in 1622, when 10 per cent of the settlers living along the James river were massacred, Smith remained confident that the Indians of both New England and Virginia could be drawn through appropriate examples and relations 'to a civil condition and thus to Christianity' (Canny 1988: 217–18).

The 'Indian massacre' of 1622, when the Powhatan Indians killed several hundred colonists in the settlements around Jamestown and destroyed their crops, marked a watershed in English relations with and attitudes to the

Indians, disillusioning many early optimists such as Smith. It is frequently suggested that the massacre rendered the Indians null in colonial eyes as inhabitants and users of their land, heralding a more full-bloodedly Lockean era of colonial ideology.[18] Yet Indian knowledge and Indian land-use remained intrinsic to colonial geography, even after the dwindling of early dependence and the violent rejection of English overtures to Christianize and assimilate the Indian. We can see this in a further geography, this time of New England, which post-dates the massacre of 1622.

Like John Smith, the Puritan settler William Wood produced a jointly verbal and cartographic geography of that part of the new world of which he had direct experience.[19] Wood's *New England's Prospect* (1634) is billed as '[a] true, lively, and experimental description of that part of *America*, commonly called New England: discovering the state of that countrie, both as it stands to our new-come *English* Planters, and to the old Native Inhabitants' (Wood 1634: n.p.).

Wood's geography has produced sharply divergent readings over the past two decades. Brian Harley deems the separate verbal accounts of native and English life in Wood's *New England's Prospect* and the 'coded language' of Wood's map, with its separate signs for Indian and English settlements, a kind of semiotic 'apartheid': part of the ideological arsenal through which Indian geography was systematically disciplined and ultimately erased (Harley 2001: 185). Folk historian Peter Benes suggests, on the other hand, that 'Wood's depiction and naming of Massachusetts Bay and the north shore ... – particularly the placing of the Indian villages "John Sagamore," "James Sagamore," ... – illustrate the closeness of English and Algonquian life' (Benes 1981: 8).

I think Wood's public geography of New England works very much like John Smith's, both to promote and celebrate the English improvement of America, and to suggest the compatibility of this project with customary native life. Wood presents New England as full of thriving, neatly nucleated settlements analogous to feudal manors, each of whose situations and qualities he describes, and abundantly stored with the commodities which will support their further growth: 'The next commoditie the land affords, is good store of Woods, and that not onely for the building of ships, and houses, and mills' (Wood 1634: 15). Nonetheless, as in Smith's writing this celebration of settlement and improvement does not prevent Wood from acknowledging a complementary Indian geography and improvement of the land which sits uncomfortably with a colonial ideology of *terra nullius* and a geographic practice of erasure.

Like Smith, Wood acknowledges, and indeed seems to celebrate, the effects of Indian husbandry. It is, writes Wood 'the custome of the *Indians* to burne the Wood in *November* ... [so that] in those places where the Indians inhabit, there is scarce a bush or bramble, or any cumbersome underwood, to be seene in the more Champion ground' (Wood 1634: 15). As in Smith's geography, this mastery of the soil is set within a complex political landscape: 'The Country as it is in relation to the *Indians*, is divided as it were into Shires, every severall division being swayed by a severall king' (ibid.: 56). Indeed the plantation of Saugus, records Wood, invoking his own home-settlement, owes its title to 'One *Blacke William*, an *Indian* Duke, [who] out of his generosity gave this place in generall to this plantation of Saugus, so that no other can appropriate it to himselfe' (ibid.: 41). And like Smith, finally, the cartographer Wood acknowledges the intimacy of the Indian with his land that naturally attends this patchwork of political custom and agrarian use. The Indians, he writes, 'are as well acquainted with the craggy mountaines, and the pleasant vales ... and can distinguish them by their names as perfectly ... as the experienced citizen knows how to find out Cheap-side crosse' (ibid.: 71).

Notwithstanding other evidence for what we understand to be a rising Lockean doctrine of improvement, rhetorics promoting settlement in America remained, like contemporary rhetorics of English surveying and improvement, necessarily flexible and heterogeneous throughout the seventeenth century. Working not to erase the customary geography of American 'common' land, but to negotiate between this customary geography and the novelties of English settlement, they responded to an environment of sustained scepticism and direct challenge.

In 1635–36 Rhode Island nonconformist Roger Williams mounted one of the most famous challenges to English title in America, posing questions which resonated throughout the seventeenth century. Williams argued that the Indians held just title to their land, which had not been superseded by the King's patent. His was by no means an unprecedented voice.[20] Moreover, objections to colonization involved more than a moral anxiety about the Indians and their rival rights. The perception persisted throughout the seventeenth century that colonies threatened the moral and material economies of the Commonwealth, diverting the attention of the nation's guardians to the mirage of foreign gold when it should be focused on the stewardship of their own immediate charges (Armitage 1998: 109–10).

In the face of this range of moral and economic objections to colonization, arguments in favour of colonization, and the geographies with which they were (often legally) entangled rarely used a proto-Lockean labour theory of value to clear the soil. More frequently they acknowledged Indian sovereignty, advancing English title incrementally, as David Armitage suggests, through a mixture of purchase, conquest and limited improvement. As well as asserting the emptiness of America, John Winthrop advanced the far more credible and far more typical claim that there was enough land in America for everyone, both native and settler, backing up his claim with biblical precedents such as Abraham, who went to live amongst the Sodomites, taking up their waste lands:

> That which is common to all is proper to none. This savage people ruleth over many lands without title or property; for they inclose no ground, neither have they cattell to maintayne it, but remove their

dwellings as they have occasion, or as they can prevail against their neighbours. And why may not Christians have liberty to go and dwell amongst them in their waste lands and woods (leaving them such places as they have manured for their corne) as lawfully as Abraham did among the Sodomites?

(Winthrop 1846: 277)

In this aspect of his argument, as in so may others, Winthrop was echoed by Locke himself. Previous readings of Locke's *Two Treatises*, notably James Tully's, have found in them a definition of civil society and property constructed in opposition to Amerindian rights, and thereby in tension with the Crown view that 'the aboriginal peoples of North America are sovereign, self-governing nations with exclusive jurisdiction over and ownership of their territories' (quoted in Arneil 1999: 107). Barbara Arneil, however, argues that Locke agreed with the Crown view that colonial title must be sought by peaceful means rather than conquest, and must acknowledge and co-exist with the title of native peoples. The Indian's right to property in America was simply limited, like the settler's, to that land which he improved. The experience of explorers and settlers in America had demonstrated, so Locke and his predecessors argued, that the Indian's very limited improvement of American land left space for their own moderate improvements.

I think we can read the public maps of early America as further evidence of the persistently ambivalent nature of seventeenth-century geography. The most characteristic function of geography in this period was not to impose Baconian discipline and dominion, but to negotiate the values and identities of contested ventures from the improvement of English estates to the settlement of America. Except in the broadest of rhetorical flourishes geographers were not inclined to clear the soil as critics from Brian Harley onward have suggested that they do. Instead they tended to present customary users, however cynically and instrumentally, as consensual partners in the improvement of English and American land.

6 Points mean prizes

Self-fashioning and the mathematical career

Like me, David Armitage has sought to set limits to the consensus which currently prevails around the imperial ethos of seventeenth-century England (Armitage 1998). Armitage has argued that this ethos was exaggerated by the humanistically trained secretaries whose work we often read: men whose primary impulse may have been to sell their own services to the ventures they promoted. All too often this pumped-up rhetoric of empire maps onto nothing more than – to use Jeffrey Knapp's resonant phrase – 'an empire nowhere' (Knapp 1992). The same could be said of the promotion of mathematics and geography themselves, whose significance in writing vastly outweighs their influence in use.

Armitage's argument chimes with a 1994 essay titled 'Pragmatic readers', co-authored by William Sherman and Lisa Jardine. Sherman and Jardine identify a distinctive form of relationship between noble employers and scholars employed as 'professional reader[s]' which they associate particularly with the 'intellectual and political life' of 1580s and 1590s England (Jardine and Sherman 1994: 102–3). Within these 'scholarly service' relationships figures such as Dee, Francis Bacon and Gabriel Harvey 'bargained' for patronage, enhancing their 'credit' with their patrons by positioning themselves as mediators of scholarly knowledge. Such mediation took various forms: sometimes the procuring of rare books; sometimes the provision of a 'reading route' through their own or others' difficult or compendious works. In all of these 'knowledge transactions' the crucial thing for the scholar was to emphasize the importance of the mediation itself, and thus the value of their services. The professional readers discussed by Jardine and Sherman positioned themselves not only as better able than other such readers to supply their masters with material and explanation, but also as better able than their masters would be on their own to navigate the sea of knowledge. This self-positioning between a patron and his books was a delicate matter, not least because the intellectual intermediacy of the professional reader was simultaneously social. Scholarly service 'lay somewhere between rank-equal friendship and servant's hire', and was generally rewarded not with wages as such, but with the more numinous rewards of social credit within a powerful and wealthy sphere (Jardine and Sherman 1994: 108).

I have no interest in proving the material importance of seventeenth-century English empire in this book or even the instrumental impact of mathematics and cartography. I'm more concerned with precisely the rhetoric of mediation whose material significance Armitage dismisses. In the last three chapters I have argued that this rhetoric did crucial cultural work in negotiating the legitimacy of artful economic endeavour in the seventeenth century, both at home and abroad. The second part of my argument in this book concerns the work this rhetoric did in negotiating a social role for mathematicians themselves. In this chapter and my final one I want to explore the way in which seventeenth-century mathematicians defined themselves and forged their careers in terms of the unstable currency of their art.

Writing and the trade in knowledge

This book encourages the perception of an ambiguous mathematics, whose unstable cultural currency was exploited as a means of cultural communication and negotiation. My approach runs counter to the impulse of historians to look variously for a determining idealist or pragmatic ethos behind early modern science.

The new social history of science makes much of its utilitarianism: what Lesley Cormack calls the 'engaged approach' (Cormack 1991: 641). It is the great Baconian vision of a renovated relationship between theoretical scholarship and pragmatic, public use that historians such as Cormack regard as driving forward such institutional reform of mathematical education as took place in the seventeenth century, which can be summarized as follows.

Whilst Cambridge was slow to acknowledge mathematics as an independent subject, not establishing the Lucasian Chair until 1662, Oxford established the Savilian Chair of geometry in 1619. Cormack has shown that university libraries began during this period to acquire an increasing number of practical mathematical texts, oriented towards such respectable patrician and mercantile interests as fortification and navigation (Cormack 1997). Within the limits of the quadrivium, students were evidently encouraged by their tutors to cultivate an interest in practical mathematics. At the vanguard of this developing mathematical culture elite circles with an experimental bent formed around mathematicians at the universities, notably the circle of John Wilkins at Wadham during the Commonwealth, which included Robert Boyle, Robert Hooke and Christopher Wren as well as John Wallis and Seth Ward, and would ultimately form the basis for the Royal Society. Outside the universities, individual schools began in the seventeenth century to loosen the grip of the Latin curriculum established under Edward VI, and in some instances to accommodate a particular emphasis on mathematics. Certain public schools chose to expand the element of their quadrivial curriculum dealing with practical mathematics, characteristically for less wealthy students and those from mercantile backgrounds.[1] Beyond the established schools and the hegemony of the liberal arts, attempts were made in the seventeenth century to make possible a more specialist mathematical education. Lectures were established in the City in 1597 at Gresham College, with the express purpose of bringing technically oriented knowledge to a 'third university', 'within reach of the ordinary Londoner' (E.G.R. Taylor 1954: 50).[2] Gresham employed a series of prestigious professors in the various branches of applied mathematics including founder members of the Royal Society. Shorter-lived were other, similar attempts to support a third, pragmatic university of mathematicians.

But this institutional evidence for the civic and pragmatic impulse behind the rise of seventeenth-century mathematics is all quite isolated and limited in a century characterized in general by a residually Aristotelian attitude to knowledge (see Webster 1975: 118–22). Where Hartlib and his peers had urged the scholar out of his study and into the fields and work-shops, the Restoration saw a 'reaffirmation of education as the intellectual formation of the courtier' (Rattansi 1972: 28). Restoration satires of the Royal Society sought to divert its members away from natural philosophy and back to the proper, gentlemanly pursuits of literature and moral philo-sophy, and the Royal Society suffered from its founding from the perception that it achieved nothing more useful than the weighing of air (Rattansi 1972: 29). In mathematical education, notes Katherine Hill, 'the general attractions of a more theoretical perspective ... grew more pronounced as a reaction to educational reformers in the 1640s and 1650s' (Katherine Hill 1998: 272). Moreover seventeenth-century writers and modern historians alike have been inclined to treat even Gresham sceptically in its aspirations to be more than a club for scientific virtuosi.[3] Perhaps most significantly of all, the quadrivium, with its implicitly idealist priority of speculative over practical, survived as the basis for most school and university education in mathematics throughout the seventeenth century, defended vigorously by such university scholars as the Oxford triumvirate of Wilkins, Ward and Wallis (Howson 1982).

More concrete evidence for the advancing status of practical mathematics in the seventeenth century lies beyond institutional education, and in the lives of individual mathematicians who succeeded, as many of their prede-cessors had failed, in making careers as mathematical practitioners. As Lisa Jardine has shown in her studies of Christopher Wren and Robert Hooke, the growing status of mathematics in the seventeenth century can be charted not just in its emergence as a valued and desired subject in both school and university education, but in the lives of prominent scientists who were initi-ated through these schools and universities into the social circles which would make possible their careers. When Robert Hooke left Busby's West-minster in 1653 for Oxford, he joined the coterie of John Wilkins, divine, mathematician and warden of Wadham, whose rooms 'were the centre of a circle of hand-picked, scientifically, mechanically and medically gifted young men', including Wren, and in the main Cavalier in politics and

Anglican in religion, 'who met there weekly to debate topical scientific issues and to carry out experiments' (Jardine 2003: 69). Hooke won prestige within this circle as a mathematician and instrument maker, and in this distinctly practical capacity served the group, including Wren, Wilkins and Robert Boyle, who would ultimately found the Royal Society and precipitate 'the dawn of the London scientific revolution' (ibid.: 75). As well as learning practical instrumental skills through which to service the theoretical inquiries of the Royal Society, Hooke and Wren both profited in their careers from the complementary humanist drive to make theory serve practice. Hooke was appointed Professor of Geometry at Gresham College in 1665; Wren Gresham Professor of Astronomy in 1657. Moreover – still sounder evidence of political faith in the practical utility of theoretical mathematics – both men were given leading roles in the surveying and reconstruction of London after the 1666 great fire, Wren as the nominee of the Crown; Hooke as the nominee of the City, whose financiers had relocated to Gresham after the destruction of the Royal Exchange. These appointments were part of a wider phenomenon whereby seventeenth-century mathematicians were increasingly trusted with the supervision of such large public projects.[4]

But whilst these public careers demonstrate support and sponsorship from elite circles in the court, the city and the universities, they still tell us little about the wider public view of mathematics. Jardine portrays Hooke's exemplary mathematical career after 1666 as torn between the practical and public ends of his City-sponsored work, and the purer scholarly ends of his service to the Royal Society, as curator of a wide array of experiments often eyed with scepticism for their lack of practical public benefit (Jardine 2003: 111). Most men who made their careers in mathematics could not afford to do so through the support of noble virtuosi who needed no persuasion of the scientific truth of mathematics, or through ongoing public employment by bodies who needed no persuasion of its utility.

Beyond the intellectual achievements of the elite groups associated with the London scientific revolution, the rising status of mathematics in the English seventeenth century was marked more pervasively in its growing social credit as a cultural commodity in a more diffuse seventeenth-century trade in knowledge. Seventeenth-century noblemen and gentlemen began to follow the example of their governors, making mathematicians members of their households or buying in their services in a variety of forms: as companions on foreign tours; tutors to their sons; and supervisors of building projects. And at a humbler level still, individual mathematicians without prestigious university pedigrees but with a proven record as practitioners were able in the later seventeenth century to establish schools specializing in mathematics. Perhaps most concrete proof of all of the wider-circulating currency of practical mathematics, some of these mathematicians were able to stake a significant part of their income on the trade in printed books.

Most mathematicians in the seventeenth century worked not just as experimental scientists or surveyors of grand designs, but privately, as instrument makers, surveyors, cartographers, tutors and writers. They did not sell their knowledge and their services exclusively to either educated scholars or to unlettered craftsmen, but to those sufficiently literate and ambitious to appreciate both the liberal caché attached to mathematical knowledge and the claims made for mathematical utility. Their success depended not on the established currency of mathematics, but on their own establishment of this currency. And what they were selling was rarely simply utility. Rather than appealing to an established intellectual understanding of the value of mathematics, or serving a social demand for a mathematics whose utility could be taken for granted, these writers, like their literary contemporaries, tended to construct a mathematics suited to the genre within which they wrote and the market for which it was designed: a mathematics which would sell.

A palatable mathematics

A notable example of the mathematical career that flourished largely away from elite court circles and the universities is that of William Leybourn. Leybourn makes no appearance in Lisa Jardine's densely realized map of the social context of Christopher Wren's seventeenth-century mathematics. Yet Leybourn might with justification be thought one of the most significant London mathematicians of his day. His long life and career paralleled Wren's closely, though at a distinct social remove.

Born in 1626, Leybourn worked originally with his brother Robert as a printer, based at Monkswell Street, Cripplegate (E.G.R. Taylor 1954: 230). Whilst the Leybourn brothers produced many books for writers involved in technical experimentation and reform, William Leybourn increasingly gave up his printing work to write his own books on mathematics; to teach private pupils whom he boarded at his home in Southall; and to work as a mathematical practitioner, taking part in private and public projects such as the great fire survey, and the surveying of estates forfeited in the Civil War (ibid.: 86).

Leybourn's practical work and his writing were clearly highly symbiotic in the maintenance of his career. Leybourn's books made evident his knowledge and competence in mathematics. They also directly advertised his services. *The Line of Proportion*, for instance, first published in 1667, offered a straightforward, accessible guide to the use of 'Gunter's line', a logarithmic series designed to help artisans with little mathematics compute areas and volumes mechanically. An unfussy duodecimo volume, it was first published in the immediate aftermath of the great fire and dedicated to the City grandees who oversaw the surveying of the ruins. As well as advertising Leybourn's employment in this survey, later editions of this simple, practical book were used to tout for further business with a wider clientele. One edition (1673) carries the following notice:

If any Gentleman, or other Person, desire to be instructed in any of the Sciences Mathematical, as *Arithmetick*, *Geometry*, *Astronomy*, the Use of the *Globes*, *Trigonometry*, *Navigation*, *Surveying of Land*, *Dialling*, or the like; the Author will be ready to attend them at times appointed.

Also, If any Person would have his Land, or any Ground for Building *Surveyed*, or any Edifice or Building *Measured*, either for the *Carpenters*, *Bricklayers*, *Plaisterers*, *Glaziers*, *Joyners* or *Masons* work, he is ready to perform the same either for *Master Builder* or *workman*: ...

You may hear of him where these Books are to be sold.

(Leybourn 1673: n.p.)

Self-marketing such as this is utterly characteristic of Leybourn's long and fertile publishing career. Leybourn published his first mathematical work, a treatise on surveying, in 1650. By 1682 he was able to fund production of a folio edition of his book on *Dialling* by subscription, expanding a text first published in quarto form in 1669, and furnishing it with his portrait (E.G.R. Taylor 1954: 406). An insider to the seventeenth-century print trade, Leybourn joined other early modern writers beginning to capitalize on the social currency of their knowledge; owning and selling shares in it. Having made a name for himself, Leybourn effectively franchised this name. He contributed prefaces to other writers' works, and published in many instances to promote instruments made by his associates: an early example of tie-in marketing. He was also cannily sensitive to the different markets for his books. Some, such as *The Line of Proportion*, are simply and cheaply produced, with the accent on practical use. *Cursus Mathematicus*, on the other hand, another relatively lavish folio production published in 1690 by subscription and costing twenty shillings, addresses itself to the leisured gentry. Moreover, the adaptability of Leybourn's mathematics is not limited to those elements which a 'positivist' historian of science might well deem irrelevant: the material characteristics and commercial value of his books. It also characterizes what's inside them.

In *Cursus Mathematicus* Leybourn fashions a mathematics neither scholarly nor pragmatic, but somewhere in between. This book, whose reader must be '*Mathematically* affected', looks not merely to 'agree with his *Stomach*', being profitable, but also to 'please his *Palate*' (Leybourn 1690: 47, 48). Such a palatable mathematics is neither thought nor work but recreation. Leybourn intends his *Cursus Mathematicus* for 'a dull solitude or vacancy of Business' as 'a *delightful* and *innocent expence* of vacant Hours' (ibid.: sig.A5r).

Leybourn's recreational mathematics was by no means unusual in seventeenth-century scientific discourse. The radically pragmatic humanism which promoted mathematics as necessary to human life – the humanism of Ramus, Bacon, Hartlib – was preceded and survived by a more moderate rhetoric of civic benefit. This more moderate rhetoric promoted mathematics as a natural refinement: not the possession of all, but a means by which some might better themselves. This mathematics is very much the practical mathematics of the school and university quadrivium. It is liberal in its detachment from necessary work and profit, and thereby a suitable component in the education of the gentleman. At the same time it is relatively trivial: elementary preparation for and light relief from more advanced, speculative studies.

Neither stoically unworldly nor illiberally pragmatic, this 'polite' mathematics, to use Mordechai Feingold's term, was the most commercial of intellectual products in a new trade in scientific knowledge (Feingold 1984). Its principal customers were those who considered themselves gentlemen, and marked themselves out as such by subscribing to a pricey volume such as Leybourn's *Cursus Mathematicus*, or employing a tutor such as Leybourn to teach their sons. Katherine Hill's study of a 1632 dispute between the mathematicians William Oughtred and Richard Delamain finds Delamain distinguishing his own emphasis on instrumental practice as more appealing than Oughtred's 'tedious regular demonstrations' to a gentry whose 'weighty occupations did not allow them time to study theory' (Katherine Hill 1998: 258). Delamain's rhetoric is strikingly similar to Leybourn's in his construction of a 'palatable' mathematics: '[Theory] to an inexperienced palate like bitter pills is sweetened over, and made pleasant with an Instrumentall compendious facilitie, and made to go downe more readily, and yet to maintain the same vertue and working' (quoted in Katherine Hill 1998: 258). This palatable mathematics, concludes Hill, much as Jesseph concludes of the Oxford triumvirate in their dispute with Oughtred, was designed by Delamain to secure his own place as a 'vulgar teacher' in the profitable market in mathematical services to the court and gentry, and to elbow Oughtred out.

Diamond rings and coal-pits: rhetorics of virtue and profit

I think it would be reductive to define William Leybourn's mathematics as definitively pragmatic, as some revisionist social history is inclined to do for far more scholarly figures in its keenness to de-Platonize the Renaissance. Certainly Leybourn was a pragmatic man, who promoted mathematical science in order to build his own successful career. But what he was selling wasn't simply practical utility. No less than Dee, Leybourn recognized that the most valuable commodity in a trade in intellectual knowledge was one which somehow blended practical use and profit with the truth and virtue traditionally associated with liberal disinterest. Produced for a readership who appreciated appeals to something mid-way between their stomachs and their minds, Leybourn's more expensive texts artfully equivocate between a practical and a liberal mathematics. And this market-sensitive equivocation is characteristic of a wide spectrum of mathematical writing, whether emanating from elite university circles, or the ostensibly more pragmatic third university of London.

In the writings of Thomas Hood, disciple of Peter Ramus and self-proclaimed popularizer of mathematics, we can explore the work of an equivocating mathematical rhetoric within a wider social strategy of self-fashioning. Hood, a fellow of Trinity College, Cambridge, had been appointed mathematical lecturer to the City of London in 1598, under the auspices of the London merchant Sir Thomas Smith and Lord Lumley (E.G.R. Taylor 1954: 179). He gave his first lecture at Smith's house in Gracechurch Street, and subsequent lectures at the Stapler's Chapel in Leadenhall Street. Here he sold books published to support his lectures, including, in 1590, a translation of Ramus's *Geometria*. After four years the lectures were discontinued, perhaps, supposes E.G.R. Taylor, 'because they were too academic for a majority of listeners' (ibid.: 179).

Hood's lectures were clearly part of that third university established in late sixteenth-century London, where knowledge was designed to be shared between university scholars and practical men. Their audience was an open one, composed in part of the militia got up to counter the Armada (Hooykaas 1958: 117). In a letter to Lord Burghley, Hood refers to himself as ' "Mathematical Lecturer to the Captains of the Trained Bands" ' (quoted in E.G.R. Taylor 1954: 328). His mathematical writing also had another audience. Hood was very much a client of the city grandees, his appointment a response to the pressures applied before and after him by figures from Dee to Wilkins for teaching suited to the common man.

His position tenuous, and ultimately untenable, Hood was obliged constantly to address and please both of these audiences: the students of his lectures, and perhaps more importantly their noble sponsors. In fact we can connect Hood's mathematical publications directly with his grip on employment. His first lecture, given at Thomas Smith's house and immediately published, secured him his first two years; his translation of Ramus thanks his patrons for their underwriting of a further two. In this writing produced to support his appointment Hood presented his mixed audience with a geometry equivocal between humanist pragmatism and scholarly idealism, hoping to convince both audiences simultaneously of the utility and of the traditional liberal virtue of what he taught. Like William Cuningham 30 years earlier, he placed his mathematics in both the study and the marketplace, and sought to weave a Daedalean course between the two.

It is appropriate, given its pragmatic circumstances, that Hood's translation of Ramus reproduces the Ramist dictum that geometry is 'the Arte of measuring well' (Ramus 1590: 1). The speech made and published by Hood in 1588 to inaugurate his lectures, in correspondingly humanist voice, determines the 'subject matter' of the mathematical arts as 'the world' (Hood 1974: 22). However, parts of this speech also take on a Platonist tone reminiscent of John Dee's 'Praeface'. Hood expresses Dee's distaste for 'clod and turff', exhorting his audience not to be 'looking grovelling on the ground as senceless beastes', but upward, and beyond the world's mere 'outward frame' (ibid.: sig.A3v). He plays his audience, appealing alternately to their needs and greeds as merchants and practical men, and their intellectual vanity as scholars. 'If you your selves were not Merchant men', he prompts, 'I would tel you what proffit you reape heerby, but your dayly experience saveth me that labour' (ibid.: sig.B3v). Elsewhere, conceding that they may be impatient to see in mathematics 'some use, or some commoditie fit for a common weale', Hood mocks the same utilitarian notion of profit (ibid.: sig.B1v). 'What profit hath the Diamond on your finger, for which you gave an hundred pound?' he teases: 'You can not answer me at a blush, except you aleadge your pleasure alone' (ibid.: sig.B1v). 'I would not have you measure eche thing by the price', Hood chides, 'for commonlye schollers are not covetous men' (ibid.: sig.B2v).

To reinforce this fantasy not just of profit but of liberal scholarship, Hood invokes two classical figures. The first is Aristippus, a Greek philosopher who, when shipwrecked on the shore of Rhodes, was relieved to find the 'signs of man' in geometric figures sketched upon the shore: figures which Hood calls man's 'footsteppes' (Hood 1974: sig.B3v). This allegory of a sandy geometry made as naturally as a footprint reinforces Hood's Ramist message that the life of man without geometry is barely human: 'fitter for a stie, then for a studie' (ibid.: sig.B3v). But the second classical figure invoked by Hood suggests a more unworldly value. '*Archimedes* sifting out the Goldesmithe's deceipte in making of King *Hiero* his Crown,' recalls Hood, 'could not contain him self for joy, but ran as he was stark naked through the street' (ibid.: sig.B3v).

Much like Cuningham in those figures with which I began this book, Hood tells his audience that geometry is both so worldly, so definitive of the human state that they barely need to learn it; and at the same time so precious, so sublime, that it will elevate them far beyond other men. Hood is scrabbling to hold on to his job; to patronage and 'wages'; and trying to convince his sponsors and his students simultaneously of the precious value and the plain utility of his subject. The same equivocation characterizes the mathematical writing of a figure superficially far more secure.

John Wilkins, a key figure for Lisa Jardine as the lynchpin of that Oxford circle of Commonwealth virtuosi who went on to form the Royal Society, published in 1648 a treatise titled *Mathematical Magick*. Wilkins is well remembered for his place in seventeenth-century intellectual history: he was a theorist of language and communication as well as of mathematics and cosmology. He was also extraordinary successful at navigating the choppy seas of mid-seventeenth-century patronage and place. *Mathematicall Magick* was produced while Wilkins was chaplain to Prince Charles Louis, nephew of Charles I and elector palatine of the Rhine. Wilkins dedicated the treatise to his employer, an admirer of his mathematical abilities (Shapiro 1969: 21). In the year it was published Wilkins became warden of Wadham. In this post he enjoyed not only the support of Cromwell, whose widowed sister he married in 1656, but also the trust of royalists whose sons he educated, and he was a vigorous campaigner for the independence of the university. His advancement was only very temporarily hindered by the Restoration, and he secured a series of prominent clerical posts in the early 1660s as well as helping to found the Royal Society. In the light of this illustrious and most prudently managed career, it makes sense that Wilkins used his writing much as he used the collection of mathematical marvels kept in his university rooms and shown off to illustrious guests: not simply to teach mathematical knowledge and technique, but to promote its social currency, and to construct the mathematician himself as an indispensable broker between mystical powers and pragmatic uses.

Wilkins's treatise, like Hood's speech, is a pattern of the ambivalence of early modern mathematics. It makes considerable show of a generous and altruistic contribution to the Commonwealth. Like Peter Ramus, Wilkins attributes the success of Germany in 'mechanicall inventions' to the institution of public lectures in the vernacular, suited to the 'capacity of every unlettered ingenious Artificer' (Wilkins 1648: A4v). He demands 'the reality and substance of publike benefit, before the shadows of some retired speculation' (ibid.: 7). And, Wilkins reminds his reader, 'there is also much *real benefit* to be learned' from mathematics, 'particularly for such Gentlemen as employ their estates in those chargeable adventures of Drayning, Mines, Cole-pits, &c.' (ibid.: sig.A4v). This characteristic humanist insistence that theory must serve practice is complemented by the equally characteristic suggestion that it learn from it. Wilkins tells a tale of the ancient philosopher Heraclitus, whose students are shocked to find him in a tradesman's shop, taking careful note of the activity there. Wilkins draws the moral from Heraclitus's *apologia* that plenty of 'divine power and wisdome' can be found in 'those common arts, which are so much despised' (ibid.: sig.A4r). 'Though the manuall exercise and practise of them be esteemed ignoble, yet the study of their generall causes and principles, cannot be prejudiciall to any other (though the most sacred) profession' (ibid.: sig.A4r).

But however close he encourages his theoretician to come to practise, Wilkins's rhetoric holds him distinctly clear. Heraclitus is in the craftsman's shop purely to observe, certainly not to take part. And what he observes, 'generall causes and principles,' are not what the craftsmen themselves see, however they may unconsciously enact them, but are precisely those a priori constituents of liberal science that have always constituted the scholar's domain. In fact Wilkins preserves a careful division of labour between theory and practice in mathematically derivate art. Such art may, despite its investment in the empirical and the particular, *'properly be styled liberall,'* since its foundation is independent of its superstructure, defined by Wilkins in terms of illiberal 'bodily exercise' (Wilkins 1648: 8). Accordingly, Wilkins divides his own art of mechanics into 'a twofold kind': '1. *Rationall* / 2. *Cheirurgicall*' (ibid.: 9). The rational part of mechanics, states Wilkins, deals with 'principles and fundamental notions', and 'may properly be styled *liberall*, as justly deserving the prosecution of an ingenuous [sic] minde' (ibid.: 9). 'The *Cheirurgicall* or *Manuall*', on the other hand, 'doth refer to the making of these instruments, and the exercising of such particular experiments' (ibid.: 9).

In addition to this internal division of manual and intellectual labour, Wilkins also disassociates his mathematics as a whole from real work, describing them as '*diversions* ... composed ... at my spare howers in the University' (Wilkins 1648: sig.A3r). His subject, what he calls '*mixed Mathematicks*', has been chosen 'as being for the *pleasure* of it, more proper for recreation, and for the *facility* more sutable to my abilities and leisure' (ibid.: sig.A3r). Like Leybourn's *Cursus Mathematicus*, this is still the practical mathematics of the quadrivium: a playful diversion from more serious, speculative study, designed to tickle the palate of his patron.

Finally, Wilkins fashions the mathematician and his mathematics not just in terms of genteel amateurism, but even in terms of the very 'retired speculation' he condemns, celebrating, in Platonic style, the 'divine power and wisdome' inherent in geometry (Wilkins 1648: sig.A4r). This is particularly evident where, like Hood, he uses Archimedes to promote his art. 'And then besides,' he writes:

> it may be another encouragement to consider the pleasure of such speculations, which doe ravish and sublime the thoughts with more cleare angelicall contentments. *Archimedes* was generally so taken up in the delight of these mathematicall studies ... that he forgot both his meat and drink, and other necessities of nature; nay, that he neglected the saving of his life, when that rude soldier in the pride and hast of victory, would not give him leisure to finish his demonstration. What a ravishment was that, when having found out the way to measure *Hiero*'s Crown, he leaped out of his Bath, and (as if he were suddenly possest) ran naked up and down crying *Eureka Eureka*!
>
> (Wilkins 1648: 293–4)

Allegory and ambivalence: Archimedes' death and Aristippus's shipwreck

John Wilkins divided *Mathematicall Magick* into two books: one titled 'Archimedes', representing the 'powers' of mathematical mechanics; the other titled 'Daedalus', representing mechanical 'motions'. I think Archimedes served early modern writers much as Daedalus did, constituting a figurative medium through which to explore and negotiate the unstable meaning and value of mathematics; of science; and of artfulness more generally. Through figures such as these, mathematics enjoyed a currency and a purchase in early modern culture that did not depend on the circulation of actual scientific knowledge. By invoking such figures Hood and Wilkins traded upon a set of images and anecdotes which most of their audience and readership might have been expected to know, however ignorant they may have been of geometry 'itself'. And through these figures they negotiated their own equivocal identities and social agencies as mathematicians.

As with Daedalus, the raw material of stories about Archimedes emanating from the classical tradition establishes the formal possibilities for interpretation. The Archimedes known best to the English seventeenth century originated principally in Plutarch's *Lives of the Famous Grecians and Romans*, Englished in 1579 by Sir Thomas North. Plutarch deals with Archimedes as part of his account of the Roman General Marcellus, whose attack on the Sicilian town of Syracuse was repelled for a considerable time by Archimedes' ingenious mechanical inventions. What is remarkable about Plutarch's account, and what proves memorable for those early modern writers who derive images and ideas from it, is the care with which it constructs an Archimedes who is vastly effective in the world yet somehow also not of it. On the one hand, Archimedes is the hero of Syracuse: a colossus of the battlefield. So powerful and ingenious were his engines that the Roman soldiers virtually refused to fight (Plutarch 1595: 336). And yet at the same time that Archimedes' designs are the basis for a God-like power in the world, his mind is also very much elsewhere. This is nowhere more evident than at the point at which Marcellus's siege succeeds:

> nothing greeued *Marcellus* more, then the losse of *Archimedes*. Who being in his studie when the city was taken, busily seeking out by himselfe the demonstration of some Geometricall proposition which he had drawne in figure, and so earnestly occupied therein, as he neither saw nor heard any noise of enemies that ran vp and downe the city, and much lesse knew it was taken: he wondered when he saw a souldier by him, that bad him go with him to *Marcellus*. Notwithstanding, he spake to the souldier, and bad him tary vntill he had done his conclusion, & brought it to demonstration: but the souldier being angry with his answer, drew out his sword, and killed him.
>
> (Plutarch 1595: 338)

The nature of Archimedes' remarkable detachment from this pressing material contingency is traditionally Platonic. Plutarch locates the historical origins of the mechanic arts practised by Archimedes in the desires of Greek mathematicians to illustrate intricate theorems. He also, however, notes the contempt expressed by Plato for any such 'vile and base handie worke of man' (Plutarch 1595: 335). 'Since that time,' Plutarch concludes, 'handie craft, or the art of engines, came to be separated from Geometry, and being long time despised by the Philosophers, it came to be one of the warlike artes' (ibid.: 335). Like John Wilkins, Plutarch's Archimedes carefully maintains this Platonic distinction between essential science and dispensable application, and scorns to write about his mechanical inventions, whose 'necessitie' and 'common commoditie' rendered them 'vile, beggerly, and mercenarie drosse' (ibid.: 336). This disdain for 'necessitie' conditions a handful of other stories told of Archimedes, some of which Plutarch recalls:

> And therfore that methinks is like enough to be true, which they write of him: that he was so rauished and drunke with the sweet intisements of this syren, which as it were lay continually with him, as he forgat his meat & drink, & was carelesse otherwise of himselfe, that oftentimes his seruants got him against his will to the bathes, to wash and annoint him: & yet being there, he would euer be drawing out of the Geometri-call figures, euen in the very imbers of the chimney. And while they were annointing of him with oiles and sweet sauors, with his finger he did draw lines vpon his naked body: so far was he taken from himselfe, & brought into an extasie or traunce, with the delight he had in the study of Geometry.
>
> (Plutarch 1595: 337)

Why, then, if he so disdained the vulgar needs of life, did Archimedes so assiduously and effectively practise them? Curiously, Plutarch gives us more than one answer to this question. We are told that Archimedes' engines were 'were but his recreations of Geometrie, and things done to passe the time with' (ibid.: 334). Then we are told that these engines were produced at the solicitation of his friend and relative king *Hieron*:

> who had prayed him to call to minde a little, his Geometricall specu-lation, and to applie it to things corporall and sencible, and to make the reason of it demonstratiue, and plaine, to the vnderstanding of the common people by experiments, and to the benefit and commoditie of vse.
>
> (Plutarch 1595: 334–5)

Then again, it is implied that Archimedes himself solicited the notice of Hieron, 'boasting' of the miracles he could achieve with apparatuses designed according to mechanical principles (Plutarch 1595: 335). This boastful pride is evident again when Marcellus appears, deploying his own siege engines: '*Archimedes* made light accompt of all his deuises [Marcel-lus's], as indeede they were nothing comparable to the engines himselfe had inuented' (ibid.: 334). Notwithstanding his contempt for 'vulgar needs'; for 'common commoditie'; Archimedes, it seems, is rather more attached to his inventions than might at first appear, and rather more proud of the worldly reputation he achieves by them. He is even figured paying court to a King, who rewards his ingenuity with patronage. Moreover, Plutarch makes clear to us the portability of the social credit Archimedes possessed. Had Archimedes survived the capture of Syracuse, it seems, he would have pros-pered. 'It is most true,' writes Plutarch, 'that Marcellus was maruellous sory for his death, and euer after hated the villaine that slue him, as a cursed and execrable person: and how he made also maruellous much afterwards of Archimedes kinsemen for his sake' (ibid.: 338).

W.R. Laird has explored the shifting currency in the European Renaissance not just of Archimedes' mathematical works, but of his reputation (Laird 1991: 630). Laird establishes a chronology for Archimedes' reputation from the Middle Ages to the sixteenth century. He suggests that whilst Medieval scholars valued Archimedes most for the philosophical detachment which they found symbolized in his death, the Italian humanists, in accordance with 'their emphasis on the useful', valued him as a 'practical artificer', dubbing the Florentine architect Brunelleschi, for instance, a 'second Archimedes' (Laird 1991: 631). The recovery of Plutarch, suggests Laird, translated into Latin in 1470 and thereafter vastly popular, 'established conclusively' this pragmatic spin (ibid.: 633):

> Despite the rehearsal of Archimedes' contempt for practical mechanics and his absorption in theoretical mathematics that cost him his life, Plutarch's account, centered as it was on the deeds of Marcellus and Archimedes' success at repelling his assault, emphasizes Archimedes' practical achievements as a designer of machines and assured that this would be the basis for his reputation in the Renaissance.
>
> (Laird 1991: 634)

Laird suggests that it was this humanist impetus in establishing the reputation of Archimedes that finally brought his theoretical work to the attention of those mathematicians who would find in it the inspiration for the sixteenth-century mathematical renaissance.

I share Laird's interest in the reputation rather than just the theories and practical uses of mathematics and the mathematician. But I disagree with his chronology of Archimedes' reputation, and in particular with his assessment of the reception of Plutarch within this chronology. I think the Medieval Archimedes persisted into the seventeenth century alongside the humanist one, reinforced precisely by Plutarch's ambivalent account. Moreover, I think it was precisely the ambivalence of this account, rather than its wholehearted espousal of scientific utility, that appealed to the humanist intellect, which characteristically and self-consciously vacillated between the active and the contemplative life. Most importantly, I think the ambiguous reputation of Archimedes was far from being merely a matter of intellectual fashion; of 'inspiration'; but itself served a wholly practical purpose, helping the professional scientist negotiate an identity for himself and his work between scholarship and craft.

Plutarch's account of Archimedes, as translated by Sir Thomas North, leaves us uncertain whether his legendary detachment was not just a bluff but part of an elaborate ploy to enhance his credit as a 'factor' of highly valued intellectual goods. Seen in this light, his relationship with Hieron becomes a delicate dance: the mathematician courting his royal relative in order that the

King might tempt him out of his contemplative trance and into the service of the state. Plutarch even makes Archimedes' geometry itself seem rather like the mediated texts produced by Jardine and Sherman's professional readers:

> For no man liuing of himselfe can deuise the demonstration of his propositions, what paine soeuer he take to seeke it: & yet straight so soone as he commeth to declare & open it, euery man then imagineth with himself he could haue found it out well enough, he can so plainly make demonstration of the thing he meaneth to shew.
>
> (Plutarch 1595: 337)

This is the perfect paradox of 'scholarly service' achieved: to lead the reader easily and plainly to their destination, but somehow also to remind them of the difficulty of their journey, and the foolishness of making it unguided. Even the death of Archimedes might be seen in this light as a self-promotional ploy gone wrong. As his city goes up in flames, Archimedes contrives that he be found, not manning the engines that ultimately failed to keep the Romans at bay, but demonstrating the detachment in which the power of his art is grounded. Marcellus himself would surely have appreciated this living emblem. The soldier he sent – ignorant of Plato – unfortunately could not.

Like Du Bartas's emblematic 'Geometrie', staring wrapt at figures in the 'sliding sand', the allegory of Archimedes' death leaves behind it unanswered questions about the nature of geometry and the geometer: did Archimedes care about his community; the practical applications of his knowledge which sustained it; or indeed his own life? Was he, on the contrary, a Platonist whose virtuous aloofness was never better emblematized than in his death? Against the grain of Laird's chronology, early modern writers who try to solve this riddle adopt as many positions on Archimedes as there are positions on geometry, reinforcing the instability and thereby flexibility of this figure as a means of mediating conflicting meanings and values.

Where Archimedes is figured 'with visage baise on ground' it is most often to celebrate his spiritual aloofness to the material world (see 'The Fourt Triumphe Called Fame', *The Triumphs of Petrarke*, Fowler 1914–40, I: 115; 'December, or old age', Farley 1638: sig.I2r). But just as often as Archimedes' unworldly Platonism is celebrated it is satirized. In places this satire is gentle and sympathetic.[5] Elsewhere, however, humanist indignation at such impolitic and un-civic diffidence is more severe (see Pettie 1576: 207). Geffrey Whitney's emblem of Archimedes (Figure 6.1) is intended to teach the moral:

> Awake from sleepe secure, when perrill doth appeare:
> No wisedome then to take our ease, and not the worst to feare.
> Still ARCHIMEDES wroughte, when foes had wonne the towne,
> And woulde not leave his worke in hande, till he was beaten downe.
>
> (Whitney 1586: 208)

Figure 6.1 Geffrey Whitney: Archimedes, in *A Choice of Emblemes and other devises for the moste parte gathered out of sundrie writers* (Leyden, 1586), 208. Reproduction courtesy of the British Library (shelfmark: 89.e.11).

Henry Peacham figures Archimedes similarly, though with a cross-staff rather than a chessboard, in a manuscript (1603) based on James I's *Basilicon Doron* (1599), one copy of which was dedicated to Prince Henry and another prepared for James himself.[6] The emblem is titled 'Studia Inoportuna' (Untimely Studies) and is accompanied by King James's advice that:

> as for the study of the other liberal arts and sciences, I would have you reasonably versed in them, but not pressing to be a pass-master in any of them: for that cannot but distract you from the points of your calling.
>
> (quoted in Shakespeare 1987: 21–2)

If these writers disagree over the value of Archimedes' death, they agree over its idealist meaning. And yet, as with the science of geometry itself, there is an opposite strain of commentary on Archimedes which celebrates or satirizes his worldly pragmatism. In places Archimedes is figured as a Hermetic magus able to move between the sundered spheres of a Platonic cosmology, making the virtues of the ideal effective in corporeal use (see 'The Sphere of Archimedes out of Claudian', Vaughan 1678: 41; 'Cursus & Ordorerum; or Art and Natvre', Hagthorpe 1623: st. 22, p. 34). But more often those who celebrate Archimedes' practical abilities are keen to disassociate them, and by extension mathematical science in general, from the taint of magic. In a note to his translation of Hugo Grotius's *Sophompaneas*, glossing the 'Magicians' on whom Pharaoh vainly called to interpret his dreams, Francis Goldsmith notes that Vossius, the dedicatee of Grotius's play, distinguishes between 'three kinds of art Magicke':

> Naturall, as when an Egge moves on a Table, because of Quick-silver put into it; or a Naile hangs in the aire, because a Load-stone is hid above and beneath it. Artificiall, when *Archimedes* burnt the Roman Ships with looking-glasses, and *Archytas* made a woodden Dove to fly. Dæmoniacall, when a man is carried a hundred miles in an houre; a Brazen head answers to any question, &c.
>
> (Grotius 1652: 59)

As often as Archimedes is blamed or celebrated in early modern writing for his idealism, he is blamed or celebrated for the worldly cunning of such artificial 'magic'. Many writers who figure a worldly rather than a disinterested Archimedes recall his marvellous artificial inventions (see 'The Philosophers Second Satyr of Saturne', Anton 1616: 12). Others, however, regard the same artful cunning not with admiration but suspicion, ranking Archimedes alongside Daedalus and Machiavelli as a touchstone of shrewdness and 'policy' (see *Hollands Leaguer*, Marmion 1875: II. i, p. 28).

The meaning of Archimedes' life and death clearly varies, over a century of usages, between idealist and pragmatic understandings of the meaning of his mathematical science, and positive and negative evaluations of such idealism or pragmatism. There is no clear chronology, for instance, marking a shift towards humanist pragmatism and away from Medieval Platonism: the instances I've given range from a 1576 humanist attack on Archimedes' disinterested asceticism to a 1678 celebration of his 'divine' Hermetic powers. What this suggests is that the figure of Archimedes remained available throughout this period, like that of Daedalus, as a cultural battlefield for competing understandings and evaluations of mathematical art. But not only a battlefield.

In a period of uncertainty about the meaning and value of mathematics and of worldly artfulness in general, the ambiguous figure of Archimedes afforded a rhetorical compromise. For a writer such as John Norden, Archimedes was available to suggest a geometry fit both for the 'mart' and 'to adorne the heart' (Norden 1931: viii). For John Wilkins, Archimedes functions similarly to reconcile liberal virtue and pragmatic profit. In Thomas Hood's mathematical speech, he shares this role with Aristippus.

As an incidental addition to his main commentary on the binding, loosing allegory of Daedalus's labyrinth, Francis Bacon finds the following 'worth the noting' of Daedalus's original banishment, and his subsequent employment away from home: 'artificers haue this prerogatiue to find enterteinment and welcome in all countries, so that exile to an excellent workman can hardly bee termed a punishment, wheareas other conditions and states of life can scarce liue out of their owne country' (Bacon 1619: 92). The last classical figure I want to consider, as a rhetorical bridge between the conflicting meanings and values of early modern science, is an exile, like Daedalus, who redeems his alienation through geometry.

In comparison to Archimedes and Daedalus, the philosopher Aristippus plays a minor role in seventeenth-century discourses of geometry. His association with the science originates in a story relayed by the Roman architect Vitruvius, whose *Ten Books of Architecture* had been widely available in various European languages since the late fifteenth century. Inigo Jones, for instance, owned a copy of Barbaro's 1567 edition (Harris *et al.* 1973: 63).

The story of Aristippus's shipwreck, told by an eminently practical mathematician, confirms of geometric science what Francis Bacon suggests of geometric artifice and what the story of Plutarch's Archimedes implies: that it confers a very portable social credit, and thus redeems the strangeness of the foreign. Says Vitruvius: '1. It is related of the Socratic philosopher Aristippus that, being shipwrecked and cast ashore on the coast of the Rhodians, he observed geometrical figures drawn thereon, and cried out to his companions: "Let us be of good cheer, for I see the traces of man"' (Vitruvius 1960: 167). Aristippus and company are entertained in Rhodes, where they engage in philosophical debate. Vitruvius continues: 'when his companions wished to return to their country, and asked him what message he wished them to carry home, [Aristippus] ... bade them say this: that children ought to be provided with property and resources of a kind that could swim with them even out of a shipwreck' (ibid.: 167). 'These,' Vitruvius observes, 'are indeed the true supports of life ... and neither Fortune's adverse gale, nor political revolution, nor ravages of war can do them any harm' (ibid.: 167). 'The man of learning,' he concludes, quoting Theophrastus, 'is ... [not] a stranger when in a foreign land ... and can fearlessly look down upon the ... accidents of fortune' (ibid.: 167).

Like allegories of Daedalus's labyrinth and Archimedes' death, the story of Aristippus's shipwreck features paradoxes which emblematize an ambivalence between idealism and the world. In the geometry which survives shipwrecks and indeed any of the accidents of fortune we might recognize that Platonic idealism which rendered Archimedes insensible to the sword-thrust which wrought his death. If this geometry is sign, or 'property', of human life in general, it is so because it proves that man himself, inherently a geometer, possesses something of immortality within him. But in the geometry which offers, in that other sense of 'property', a chance for some men to gain a peculiar advantage, we might recognize something rather more worldly. Not quite the geometry of Ramus, which is the natural property of the crudest craftsman. More the geometry of Daedalus in his more self-interested, artful, politic guise; or Archimedes, hired and admired by friends and enemies alike.

In this ambivalence between a disinterested, ideal geometry and a geometry of worldly artfulness we return, once again, to the ambiguous materiality of figures drawn upon the ground (Figure 6.2). Do these figures on the Rhodean shoreline signify scholarly transcendence of the accidents of fortune, or the craftsman's capacity to manage and indeed capitalize upon such accidents? Vacillating implicitly between the philosophical and the practical guises of mathematics, Vitruvius suggests that a traveller might travel around the world in a perfect circle, returning home exactly the same, and yet still bring with him some material benefit from abroad.

In comparison to stories of ingenious mechanics such as Archimedes and Daedalus, it might seem to push interpretation of this Aristippus allegory too far to find in it signs of worldly interest. Aristippus is, after all, a philosopher, as are the mathematicians he encounters in Rhodes. And yet what sets the reputation of Aristippus apart from the reputations of these other two geometers in the Renaissance is not his idealism, but in fact the opposite. Rarely mentioned in early modern literature in connection with geometry, Aristippus is more typically associated with his mockery of Diogenes' cynic stoicism, and with his epicurean insistence that the philosopher might fully enjoy material, physical pleasures, and in particular those derived from service to the powerful and rich. In early modern literature, these commitments to the world inspire widespread figurations of Aristippus as epitome of court duplicity and flattery, of licentiousness, and of drunkenness in particular.[7] Touching ironically upon the anecdote of geometric traces in the sand, the Puritan Stephen Gosson (1579) concludes of Aristippus that 'if we followe the print of his feete, I finde that we differ not from sauage beastes' (Gosson 1579: 68). Gosson urges the courtier to submit to learned discipline rather than succumbing to Aristippean pleasure, for:

> if courtiers begin to despise knowledge, and thrust theyr Philosophers out of the gates, all wisedome, all nurture, all good maners, al gouernment, all honour and honestie goes too wracke. Plato had not been one houre out of Dionysius fauour, but euery one of his lessons was turned too a daunsing trick, every Gentlemans Pen set a work with the prayse of his Mistresse; and euery Geometrical figure drawen in the botome of a boule of Wine.
>
> (Gosson 1579: 28–9)

In the light of the all-too-worldly footprints with which most early modern writers associate him, Aristippus's sand-drawn, even wine-drawn geometry

Figure 6.2 Michael Burghers: frontispiece in Ευκλειδου τα σωζομενα. *Euclidis quæ supersunt omnia*..., trans. David Gregory (Oxford, 1703). Reproduction courtesy of the British Library (shelfmark: 678.k.6).

seems to me just as ambiguous between 'loosing' and 'binding' as Archimedes', and just as fertile an equivocation between the study and the marketplace.

Living emblems

In encouraging his mixed audience to follow in Aristippus's footsteps, Thomas Hood was asking them to believe, in Bacon's terms, that they might loose themselves from the world as scholars even whilst they bound themselves as merchants and courtiers ever more tightly to it. Like Archimedes, in the mathematical fables whose popular currency his speech both presumed and worked to extend, he was selling the authority of liberal transcendence and the agency of worldly use. He needed his audience to share with him this fantasy of ideality merged with use; and he needed them to regard him as uniquely capable of guiding them on this numinous in-between path. No less was this true of John Wilkins, negotiating his own position within the patronage system of mid-century scientific virtuosity.

The myth of the magus, then, is far from an anachronism in our view of early modern science. Neither is it a Medieval aberration in a steadily evolving modern scientific culture. It was an integral part of the identity the mathematician fashioned for himself in his attempt to sell his services. The success of such self-mythologies may be judged in their repetition not just by the modern biographers of early modern mathematicians, but by their contemporaries.

While Robert Hooke was gaining his practical education at Westminster, notes Lisa Jardine, Christopher Wren was receiving a 'rigorous training in pure mathematics ... in the circle of the great mathematician William Oughtred' (Jardine 2003: 61). As well as Billingsley's Euclid, Hooke's own education was furthered by Oughtred's *Clavis Mathematicae*, published in 1631 (Jardine 2003: 60; Oughtred 1631). This Latin text-book was the first of a substantial corpus of works associated more or less directly with Oughtred's name. In 1632 the instrument-maker Elias Allen published a work by Oughtred's pupil William Foster titled *The Circles of Proportion and the Horizontal Instrument. Both inuented, and the vses of both written in Latine by Mr. W. O.* (Oughtred 1632). This translation of Oughtred's Latin notes, for which Allen produced accompanying instruments, was clearly a success, and was reissued at least three times, the last in 1660. In 1652 Christopher Wren, the 20-year-old acolyte of John Wilkins at Wadham, contributed to a new edition of Oughtred's *Clavis*, which again went through several editions.

From his first entry into print, Oughtred's name was clearly very much in the world, and enjoyed considerable currency in the trade in printed knowledge. For a time Oughtred's work was, as Jardine puts it, in 'fashion' on the 'mathematical scene' (Jardine 2003: 60, 62). Yet both Elias Allen and William Foster characterize Oughtred as being persuaded into print against his natural inclination, and as profiting nothing from the enterprise. It is almost as though they divide the two aspects of the ambivalent Archimedean persona, using Oughtred the disinterested philosopher, writing even his notes in Latin, as a reservoir of intellectual value with which to underwrite their practical, profitable, vernacular applications. In this division they copy Oughtred himself.

As they fought their battle of words in 1632 Oughtred and Richard Delamain staked conflicting claims for the proper relationship of theory to practice. Oughtred, a university scholar, professional clergyman and tutor to the Earl of Arundel's son, claimed that Delamain taught mere 'Juglers' tricks: instrumental practice without theory (Katherine Hill 1998: 254). Delamain, a sometime joiner with no university education turned royal client, happy to identify himself as a 'vulgar teacher' to those whose primary interest was practical, claimed that the more scholarly Oughtred embroiled his students in a 'labyrinth' of theory (ibid.: 258). Katherine Hill's first major conclusion from this dispute is that 'movement between theoretical and practical activities' in the seventeenth century 'could easily pass through a contested zone' (ibid.: 257). Her second is that where such ambiguity prevailed, mathematicians were obliged to 'negotiate' their authority and simultaneously 'the local meaning of mathematics' in relation to competitors and in relation to the perceived needs and desires of their 'potential students and patrons' (ibid.: 259, 271, 274). It's highly plausible to regard Oughtred's strange ambivalence towards the mathematical instruments and books he designed, wrote and saw marketed in his name as a highly market-sensitive exercise in self-fashioning. And whether or not he made money directly from these commodities, Oughtred surely benefited from his divided persona.

William Oughtred was the son of an Eton scrivener and writing master, and a King's scholar at Eton and Cambridge, where he remained for eleven years (E.G.R. Taylor 1954: 192). He cultivated a social role for himself that had, like John Wilkins's, far more to do with professional scholarship and patronage than with the waged services which William Leybourn was willing publicly to embrace. His first mathematical writing was produced in 1597 and circulated in manuscript. He taught at Cambridge; when he left Cambridge secured preferment at Albury near Guilford, giving one of his instruments to the bishop who ordained him; and for some years was employed as tutor to the Earl of Arundel's son. His first publication, *Clavis Mathematicae* (1631), was written ostensibly to support these lessons (Oughtred 1631). No doubt the relationship was meant to work both ways, Arundel's patronage lending status to and helping to sell the book.

The considerable success of this Latin publication, alongside his equivocal association with more practical publications, allowed Oughtred to hedge his bets between scholarly, polite and practical mathematics, much as Leybourn did across his varied corpus. It allowed him to exercise what Taylor describes as 'a formative influence on a host of young men with a mathematical bent, alike at the University level and at the instrument-maker's bench' (E.G.R.

Taylor 1954: 192). In exercising this carefully refracted social influence, Oughtred enjoyed the humanist virtue of benefiting the Commonwealth without sacrificing the liberal aloofness which rendered him and his books suitable for a gentleman's house and education. In effect, he lived the allegory of Archimedes, philosopher and reluctant courtier. This comparison seems to have impressed itself, perhaps unconsciously, on his contemporaries. Oughtred's appearance in John Aubrey's *Brief Lives* is as unworldly stoic: teaching students *gratis*, dressing in ancient clothes, and amazing 'learned foraigners' with the frugality of his life (Aubrey 1972: 384).[8] Like Archimedes, abstracted from his material surroundings in the most tawdrily material of circumstances, Aubrey's Oughtred would absent-mindedly 'drawe lines and diagrams on the dust' (ibid.: 383).

The Archimedean persona of his early teacher seems to have impressed itself on Christopher Wren himself, and to have played a part in his own self-fashioning. Robert Hooke wrote of Wren in his 1665 *Micrographia* that 'since the time of Archimedes, there scarce ever met in one Man, so great a Perfection, such a Mechanical Hand, and so philosophical a Mind' (quoted in Jardine 2002: 182). In this description of the dual, Archimedean nature of his friend, Hooke was merely furthering Wren's own emblematizing of himself. In Wren's 1657 inaugural address as Professor of Astronomy at Gresham he indicated his reluctance to leave the scholarly seclusion of the Wilkins circle at Oxford for pragmatic London and public service. Like Archimedes Wren would, he said, have preferred to 'exercise [his] radius in private dust' (quoted in Jardine 2002: 130). Presumably, like Archimedes, Wren would have been disappointed had his patrons taken him at his word.

Points mean prizes

Renaissance mathematicians clearly made no attempt to hide their ambivalence between virtue and profit. In fact they drew attention to it, and thus to their own work of mediation, particularly in their writing. These conspicuous textual negotiations were an integral part of the mathematician's self-fashioning as a mediating figure: a professional scholar and factor in knowledge. The more literary the mathematical text, the more conspicuously artful the mathematician's negotiation, and the more valuable his textual work of mediation. The problem with the 'plain' mathematical text, as mathematicians themselves sometimes complained, was that it made invisible the work involved in mathematics, and seemed to render the mathematician himself dispensable. The trick was to show the path but also sell the guide. Moreover this is true not just of the verbiage which we might suppose surrounded early modern geometry 'itself', but even of the elements which formed its apparent core. Mathematical texts do not just supplement their geometries with rhetorical embellishment; these elements themselves are often conspicuously figurative: textual negotiations demonstrating the radically cultural nature of geometry.

This is true, once again, across the spectrum which runs between a Platonic and a pragmatic-humanist geometry. Only Hobbes's ultimately discredited geometry unwaveringly materializes its first principles. Whilst it defines the ultimate objects of mathematics as ideal and utterly divorced from 'clod and turff', John Dee's preface to the first English Euclid also describes an artful process of refinement by which this gulf can be bridged:

> Though *Number*, be a thyng so Immateriall, so diuine, and aeternall: yet by degrees, by litle and litle, stretchyng forth, and applying some likenes of it, as first, to thinges Spirituall: and then, bryngyng it lower, to thynges sensibly perceiued: as of a momentanye sound iterated: then to the least thynges that may be seen, numerable: And at length, (most grossely,) to a multitude of any corporall thynges seen, or felt: and so, of these grosse and sensible thynges, we are trayned to learne a certaine Image or likenes of numbers: and to vse Arte in them.
>
> (Dee 1570: sig.*1r)

A humanist geometry such as Peter Ramus's starts from a completely different premise to Dee's: that geometry is originally derived from human use. Like Aristotle, however, Ramus and other humanist mathematicians retain the priority of abstract mathematical knowledge in their science, and so always begin, like Dee's and any other geometry, with Euclid's elements. The influential English humanist and vernacular publisher Robert Record sets his agenda in *The Pathway to Knowledg* by abandoning the indivisible, 'unsensible' point of 'onelye *Theorike* Speculacion', in favour of 'that small printe of penne, pencyle, or other instrumente, which is not moved': a definition more suitable for 'practise and outwarde worke' (Record 1551: sig.A1r).

Like Du Bartas's emblematic 'Geometrie', Record performs the ambivalent gesture of averting his gaze, to fix it on his pragmatic, pen-drawn geometric points. In his case it is the ideal he turns away from, but without denying its presence. Indeed he reminds us of what he neglects. In this ambivalent gesture Record echoes the humanist pragmatism of Florentine practical mathematics a century before. In his writings on perspective in painting, Leon Battista Alberti advises that the painter should think, for instance, of the geometric point, or '*signum*', not so much as mere signifier of the ideal Euclidean principle of indivisibility, as the 'mathematician' would, but more as 'somehow a kind of thing between the mathematical point and . . . finite particles like atoms' (quoted in Edgerton 1975: 81). Such an atomized geometry recalls once again the ambiguous materiality of the dirt-drawn figures which fascinate Du Bartas's dusky-buskined goddess and Archimedes himself.

Alberti claims to speak 'not as a mathematician but as a painter', and to express himself in 'cruder terms' than those of pure mathematics: terms which figure the mathematical line, for instance, as a 'brim' and as a 'fringe'

(Alberti 1972: 36).[9] And it is precisely these crude, painterly, metaphorical terms that allow Alberti 'the artist' to speak at all. A genuinely ideal mathematics endures no material or rhetorical figuration, and a genuinely pragmatic one no special authority raising it above the realm of mute, anonymous craft. For Dee, in his most Hermetic moments, it is the mathematical text itself which builds the artful, 'stretching' bridge between 'real' mathematics and the world. But in reality no less is true of those pragmatic texts which ostentatiously avert their gaze from 'theory' and build an inescapably textual, rhetorical middle ground between abstraction and pragmatic use. And no less is true in the case of a polite and palatable mathematics, where Record's textual geometry becomes the playful metaphorical nothingness of stage satire: the geometry of the beggar's clothes.

In *Mathematical Recreations*, a popular text traditionally ascribed to William Oughtred, the elements of geometry are self-consciously trivial.[10] Published in 1633, *Mathematical Recreations* addresses itself to an audience, the '*Nobillitie* and *Gentrie*', who look to education 'to content and satisfie their affections, in the speculation of such admirable experiments as are extracted from them', rather than 'in hope of gaine to fill their *purses*' (Oughtred 1633: sig.A1r). Rather than exclusively at the head of immutable Euclidean commandments, this text locates the first principle of geometry amongst a cache of trifles. 'Sundry fine wits,' observes the author, 'as well amongst the Ancient as Moderne, have sported and delighted themselves upon severall things of small consequence, as upon the foote of a *fly*, upon a *straw*, upon a *point*, nay upon nothing' (ibid.: sig.A2r).

No less than Dee's momentary sounds, Record's pen-drawn points, Alberti's brims and fringes, or indeed Du Bartas's figures in the sand, this polite, fly's-foot geometry exists solely on the page, in a realm of metaphor which parts the ideal from the material and advertises itself as artful, textual bridge between ideal virtue and worldly pleasure/profit. This is no Husserlian geometry, kicking away the artificial props of language to reveal the eternal absolute, but the rhetorical geometry of Derrida, conspicuously endangering itself through its entanglement in the labyrinth of writing and the world. In my next and final chapter I will show how seventeenth-century surveyors promoted and profited from this rhetorical entanglement, fashioning themselves as middle men.

7 The doubtful traveller
Surveying and the middle man

If we were to construct a phenomenological 'poetics' of the geometry characterized in the previous chapter we might conclude that its essence was the human desire to move between temporal and eternal spheres. In re-constructing such a poetics we would have to begin with the mathematical images which bridge the ideal and the material realms: Dee's, Record's, Ramus's and Alberti's tiny points and momentary sounds. We might also conclude that one of the most characteristic images of intermediacy is the study. Several recent scholars of early modern cartography have noted the close association of maps with the studies in which writers imagine them being consumed. As Garrett Sullivan neatly articulates it: 'the map ... performs the surprising function of realizing the spaces of textual consumption' (Sullivan 1998: 96). And the figure of the study, as Sullivan also notes, performs the distinctive and crucial cartographic task of eliding the distinction between representation and reality; between looking at and being in a place (ibid.: 97).

The mathematician's study might be considered an important site of contention for the various schools of a new critical history of cartography. A Foucauldian map-reading sees in it the sinister profile of the panopticon. Critics tracing the cultural poetics of cartography have been inspired by phenomenology to soften these outlines and find in the study the traces of universal human desires for transcendence, secrecy and comfort.[1] A materialist social history tends, on the other hand, to be interested in the mathematician's study principally as a concrete location of particular events and material exchanges.[2] But none of these interpretations quite explains for me the cultural work done by the figure of the study: the relationship between what is inside and what is outside the mathematician's text. The first two 'reduce' the figure of the study to clean, idealist absoluteness, and the last obscures its figurative nature.[3]

I think we need to appreciate both the rhetorical, textual nature of the study and the cultural specificity of the work this figure does. John Dee's famous study, for instance, was certainly a house at Mortlake where he read and wrote particular books and met the movers and shakers of Elizabethan London. It was very much a 'space' outside the written text and inside the materialist historian's irreducibly social world. Much the same could be said of Wilkins's famous rooms at Wadham, or of Oughtred's study at Albury, where he received a host of visitors who came to marvel both at his mind and at his famously eremitic demeanour. But studies such as these also functioned inside texts, and in the wider discourses which traversed them, to signify the intellectual commodity and the social credit gained by learning mathematics. Not just a representation of real space, they functioned as key tropes in a cultural discourse negotiating truthful and legitimate representation.

The most characteristic impulse of the seventeenth-century mathematician was neither to transcend the world nor to embrace it, but to draw attention to their own negotiations; the scholarly services they could render. This makes mathematics an inescapably rhetorical discipline, perpetually performing its fall from Icarean ideality into the world. It is with this insight in mind that I want to return to the mathematical practice of surveying. Like David Armitage I think the seventeenth-century geographies we know best were the products of concerted self-fashioners eager to promote ventures in which they might serve a part. From John Norden to John Smith these 'factors' and 'traders' in knowledge exploited the unstable values of their cultures and their arts to fashion themselves as indispensable middle men: gatekeepers of moral value and intellectual truth.

Colonel Vitruvius: the unstable currency of surveying

As with mathematics in general, we can measure the unstable seventeenth-century currency of surveying in literary discourse. Compared with the prominence of stewards, there are few actual surveyors in early modern literature. Even Robert Aylett's Joseph, whilst he 'surveys' Pharaoh's land, is never described as a surveyor, although frequently as a steward. But I think this absence is significant in itself. In a sense the entire validity of the traditional moral economy is at stake in the character of the steward, and especially so where this economy is perceived as under threat from the forces of the market and the attendant derelictions of its natural guardians. If the steward can be regarded as loyal and honest; as reliable in acting in his master's interests as the master himself; then he can reliably stand guarantor for a system founded on goodwill. Where character is so much at stake, the frisson of the bad steward and the consolation of the good are clearly rich material for drama. In the case of the surveyor the opposite is true. The surveyor constitutes his identity as surveyor in a detachment from the particular, personal concerns that constitute theatre, and since his art is founded upon abstract science, rather than personality and local knowledge, his character is supposed to be of no account at all.

If actual surveyors are generally absent from the early modern literary picture, the act of surveying is not. As with the pictures illustrating contemporary surveying manuals, the general action has simply become abstracted from the particular actor. This is particularly the case where surveying is figured in positive terms. Where Andrew Marvell writes imaginary 'Instructions to a Painter' (1667) for a coruscating representation of the state of the nation after the Dutch war of 1664–67, it is a critical detachment from the corruption and decadence of the Restoration political scene that is evoked where the painter is asked to 'rest a little, and survey / With what small arts the public game they play' (*The Last Instructions to a Painter, About the Dutch Wars 1667*, Marvell 1972: ll. 117–18, p. 160). But as with mathematics in general, not all seventeenth-century figurations of surveying are positive. In fact where surveyors themselves make a rare appearance, it is mostly in order to un-mask their claim to liberal disinterest.

Ben Jonson's *Loves Welcome at Bolsover*, an entertainment for King Charles performed in 1634 (first published 1640), begins in utterly Platonic style with a chorus praising 'Love' as a:

> *lifting of the Sense*
> *To knowledge of that pure intelligence,*
> *Wherein the* Soule *hath rest, and residence.*
> (*Loves Welcome at Bolsover*, Jonson 1925–52, VII: ll. 3–5, p. 807)

The entertainment continues with a Platonic hierarchy of the senses, from the lowest (taste) to the highest (sight). But when Colonel Vitruvius appears upon the scene, identifying himself as a 'Surveyour', it is clear that something more earthy is signified. Vitruvius speaks in bumpkin-ese, and comes attended by a band of mechanics over whom he fusses. 'Doe you / know what a Surveyour is now?' he asks the audience, answering:

> I tell you, a Supervisor! / A hard word, that; but it may be softned, and brought in, / to signifie something. An Overseer! One that oversee-eth / you. A busie man! And yet I must seeme busier then I / am.
> (Jonson 1925–52, VII: ll. 43–8, p. 809)

Jonson's surveyor touches here upon an important and problematic paradox. The steward might be thought busy, and indeed ought to be so, however much his busy-ness might be satirized on the early modern stage, and signally in the form of Shakespeare's Malvolio. But the surveyor's 'work' is of a supposedly less laborious kind: busy, and yet not so – much, presumably, to the resentment of the ordinary labourer and craftsman. Hence Colonel Vitruvius's embarrassment. And if this Vitruvius avoids work, it is not in favour of a contemplative life.

Vitruvius and his mechanics proceed to make music and dance a clumsy dance: the parodic antithesis of the dance of Daedalus performed in Jonson's *Pleasure Reconciled to Virtue*. The lame smith beats upon his anvil and the 'Plaisterer', 'Morter-man' and the rest perform with 'Tune, and Measure' (Jonson 1925–52, VII: ll. 49–62, pp. 809–10). When they have finished, Vitruvius commends them:

Well done, my Musicall, Arithmeticall, Geometricall / Gamesters! or rather my true Mathematicall Boyes! It is / carried, in number, weight, and measure, as if the Aires / were all Harmonie, and the Figures a well-tim'd Proportion!

(Jonson 1925–52, VII: ll. 67–70, p. 810)

If *Pleasure Reconciled to Virtue* suggests that a graceful, courtly geometry might reconcile the Platonic ideal with worldly pleasure, *Loves Welcome* playfully encounters and explodes the pretensions of the surveyor to comprehend anything but the most muddy form of measure.

Jonson's satire on practical mathematics is undoubtedly a product of the poet's stormy relationship with Inigo Jones. It is also a kind of anti-masque: an inversion of a proper order where such things would be taken seriously; accorded the respect they deserved. But as in general where ignorant prejudices about mathematics feature in early modern discourse from surveying treatises to the comic stage, this satire registers a concrete scepticism about the capacity of mathematics to be any more than sterile Platonism or muddy craft: a 'nothing' on which no 'something' could be hung.

In Jonson's entertainment the un-busy business of the surveyor is laughable: a confidence trick with no serious consequences. Elsewhere, the authority of the surveyor is regarded with a more serious suspicion. Thomas Middleton's *Anything For a Quiet Life* (produced 1622; first published 1662) harps once again on the theme of improvident second wives (*Anything For a Quiet Life*, Middleton 1885–86, V). The widower Sir Francis Cressingham has married a young girl of 15 whose courtly pedigree will, his friends warn him, be his ruin. Sure enough, at the beginning of act IV he is close to yielding to her insistence that he sell his estate and cashier his heir to fuel her extravagances. Saunder the steward is clearly in cahoots with his mistress, urging Cressingham 'to translate this / unnecessary land into ready money' (Middleton 1885–86, IV. i. 14–15, p. 302). A warning voice is briefly sounded by a surveyor who makes a momentary entry on the stage as the act opens. Cressingham asks him if he can predict the weather. The surveyor retorts: 'Marry the fowlest weather is, that your / Land is flying away' (ibid.: IV. i. 6–7, p. 301).

It would be convenient to extrapolate from Middleton's play the accumulating legitimacy of surveying over stewardship in contemporary culture, coded in the representation of a steward who is corruptible, and a surveyor who is not. However, the surveyor is clearly unknown to Sir Francis, and has in all probability been employed by his wife and her accomplices to measure up his assets for 'translation' into cash. Middleton's surveyor may have a perspective on the crisis of Cressingham's estate, but unlike the loyal manorial steward of many contemporary representations he cares not a jot about it. He represents the threat of the modern, the urban, the commercial, even the feminine, in a moral narrative of corruption.

Where stewards are 'inside' figures who facilitate the intimate management of day-to-day affairs, the surveyor has been brought in from outside, as Joseph was by Pharaoh, to reckon up a cool mathematical account of assets in a crisis. This mathematical dispassion is typified in another poem written on the Dutch war. Royal Surveyor-General Sir John Denham's *Directions to a Painter* (1667), like Marvell's *Last Instructions to a Painter*, is a satirical pastiche of Edmund Waller's *Instructions to a Painter* (1665), a panegyric to James, Duke of York's 1664 engagement with the Dutch fleet off Lowestoft.[4] Denham's poem berates 'an ill-govern'd State', and particularly the failure of King and parliament sufficiently to prepare the nation against the Dutch (Denham and Marvell 1667: 33). Recalling the damaging Dutch assault in 1667 upon an unprepared English fleet in the Thames, Denham pictures Charles II surveying a London in crisis with a culpable rather than virtuous detachment:

> The King, of danger now shews far more fear,
> Than he did ever to prevent it, care;
> Yet to the City doth himself convey,
> Bravely to shew he was not R[...]n away: ...
>
> As *Nero* once, with Harp in Hand, survey'd
> His flaming *Rome*; and as that burnt, he plaid:
> So our great Prince, when the Dutch *Fleet* arriv'd,
> Saw his Ships burnt; and as they burnt, he –.
> (Denham and Marvell 1667: 34)

The problem with the liberal art of surveying is that it is – by definition – open to anyone. Where the steward is an appointee, qualified by personal qualities of honesty, prudence and experience, the surveyor is a blank, whose figure gestures emptily towards abstract science. In the same poem which imagines the painter's detached survey of a Britain in decay, Marvell invokes the figure of Michiel de Ruyter, who commanded the Dutch incursion into the Thames:

> *Ruyter* the while, that had our Ocean curbed,
> Sailed now amongst our rivers undisturbed,
> Surveyed their crystal streams and banks so green
> And beauties ere this never naked seen.
> Through the vain sedge, the bashful nymphs he eyed:
> Bosoms, and all which from themselves they hide.
> (*The Last Instructions to a Painter, About the Dutch
> Wars 1667*, Marvell 1972: ll. 523–8, pp. 170–1)

Now the same masculine gaze which surveyed the effeminate luxuries of the Restoration court is raping Britain, in the guise of the more disciplined, martial and cartographically dominant Dutch.[5] Marvell casts Fulke Greville's vision of a balanced mathematical improvement, copied from the 'Hollander', in a markedly paranoid light.

Mathematical fantasy: the panoptic chair

Like Andrew McRae I think that the rhetorics of mid-century Puritan reformism transformed the meaning of artful improvement. But beyond this reformist fervour – before, alongside and after it – those mathematical practitioners whose work was entangled in the capitalization of the English economy had to face a profound contemporary scepticism about the legitimacy, truth and usefulness of their art. They stood accused of culpable detachment; self-interested cunning; and a reckless disrespect for English custom. Even the surveyors of James Harrington's perfectly balanced utopian Commonwealth of Oceana are suspected of corruption (Harrington 1992: 87). In this uncertain climate of shifting truths and values mathematical practitioners did not readily let go of the aura of moral authority, discreet experience, and personal trustworthiness associated with traditional stewardship. Whilst they made increasingly confident claims for the powers of mathematics, they also negotiated carefully between this new liberal foundation of their work, and elements which marked it out as personal and professional.

Mathematics, in this negotiation, undoubtedly came first. It was the element in surveying that lifted it above manual craft and into the dignified company of the liberal arts. Hence the prominent rhetorical attention paid in surveying manuals to the geometrical foundations of surveying. I think these manuals, often treated by cultural historians as evidence of a widespread revolution in social attitudes to the use of land, are better understood in terms of the mathematician's own self-fashioning.[6] The purpose of these texts was not just to instruct in practical techniques, but to advocate for the use of mathematics: to spread a 'fantasy' of mathematics, however hollow, and thereby to negotiate a role for the mathematical practitioner.

From the late sixteenth century onward mathematical publicists and practitioners began to push mathematical knowledge, instruments, practices and maps as part of an extended range of commodities and services. They no more restricted what they offered to what was most useful or most likely to be used than any modern salesman does, but through an exorbitant mathematics 'idealized the estate surveyor's skills and exaggerated his importance and status' (P.D.A. Harvey 1996: 31).[7] This mathematical exorbitance is exemplified not just in the exotic range of instruments encountered in practical mathematical texts, many of which were almost certainly rarely if ever used, but in their often extravagant range of techniques, and even in the primary technique of triangulation itself.

Triangulation, in a treatise oriented to practical use, was a form of mathematical showing off. Requiring a precise measurement of angles beyond most practitioners and their instruments, triangulation represented, in P.D.A Harvey's words, 'an unusually early case of practical techniques developed from abstract theory' (P.D.A. Harvey 1980: 163).[8] As such its promotion contradicts Lesley Cormack's notion of surveying as an eminently practical discipline oriented to eminently practical men. But the difficulty of applying a technique such as triangulation does not mean that it belonged more to the 'commonwealth of ideas' than to the 'marketplace.' Showing off their mathematical capacities, mathematical promoters advanced their own authority and marketability and that of their discipline as a whole.

The more the mathematician 'reduces' the particularity of their subject, the smaller the contribution of their individual discretion, and the greater their confidence and authority appears to grow. Working with pure geometry the surveyor is liberated precisely as he binds himself. John Love shows in *Geodaesia* how to 'compass in' a given plot 'with one great Square,' and then subdivide into little ones (Love 1688: 137). 'This done,' he says, 'see in what Square, and part of the same Square, any remarkable accident falls, and accordingly put it down in your lesser Squares'. The theodolite 'enforceth grounds,' states Agas, 'unto the forme, likenes and similitude of any figure in the worlde' (Agas 1596: 14). The fantasy of the perfect map clearly brings with it a fantasy of unhampered cartographic agency: the fantasy encapsulated in the persistent figure of the study.

In some waste part of the paper: discretion and the margins of the map

Like all mathematicians, surveyors like to imagine themslves not just overlooking their subject, but contemplating it from afar. Cyprian Lucar (1590) suggests his readers picture themselves 'sitting at home in your chaire' (Lucar 1590: 53). Agas sees them comfortably pondering orderly geodesic quotations which may be 'set out' 'by any scale at all times, when you please,' and examining 'evidence' 'much readier than in the fields' (Agas 1596: 13). Once made, the perfect map can afford its user a similar detachment. The lord is depicted 'sitting in his chayre at home'; 'in his study, or other private place'; viewing 'at any time'; 'at pleasure' the state of things in the field (Worsop 1582: sig.B3v; Leybourn 1653: 274, 275).

Yet if the mathematical writer sells his client a potent and omniscient mathematics, placing him on a panoptic chair, there is another aspect to the rhetorical rhythm of the surveying manual generally neglected by recent critics. Despite his rhetoric of reduction, the surveyor does not excise the non-mathematical elements of his survey; he merely pushes them to its margins, leaving space for the qualitative and the legal; for words, negotiations, and for his own indispensable stewardship. We can trace this equivocal manoeuvre in Leonard Digges's account of the whole procedure of taking a survey.

Digges's narrative constructs the meaning of the survey symbolically through a series of actions and movements analogous to those which Steven Shapin has charted across the seventeenth-century 'house of experiment' (Shapin 1988a). Digges demonstrates the process by which a subject is 'reduced' and by which the surveyor himself simultaneously retreats from the physical environment of his survey to a position of virtual transcendence, whether marked by an actual study, or by a drawing board. But he also describes a phase of cartography subsequent to mathematical reduction. When the compass marks of their plan have been erased, he instructs, surveyors should complete their map as follows: 'beautifie it with ymages and Figures, as you thinke most agreable and fitte to expresse and represente the patterne, I meane the country it selfe that you describe' (Digges and Digges 1571: sig.L3r). The return of detail denoting the discrete characteristics of a subject is clearly attended in Digges's account by the return of individual discretion in the surveying task. So it is with other writers. In his instructions on how to 'draw a fair draught' from a general survey, Love tells his reader that he should 'beautifie' the whole 'as you shall see convenient' (Love 1688: 144).

A fundamental distinction remains here between the grounding phase of the cartographic process, tightly governed by scientific method, and the surveyor's discrete, judicious additions. This hierarchy is irreversible. It is marked in the familiar logic of the parergon, whether embodied in such spatial metaphors as that of adornment, or in the physical spatiality of cartographic margins. Where Leybourn tells how to 'furnish'; to 'trick and beautifie' his map; William Folkingham includes a section on 'the Tricking of Plots and Maps with Colours, Characters, Courts and other Complements'; and Love instructs his reader to place his 'convenient' additions 'in some wast part of the Paper' (Leybourn 1653: 174; Folkingham 1610: sig.A1v; Love 1688: 144).

If it stands in for discrete, local information on the map, the graphic parergon also may be said to stand in for words, narrative; even history. Sir Balthazar Gerbier, in his *Publick Lecture of an Introduction to Geographie* (1649), divides geography Ptolemaically into 'Geometricall' and 'Historicall' parts, and attributes to the latter responsibility for conveying the 'proprieties of every Region' (Gerbier 1649: 2). History, like pictorial parerga, casts a shadow not allowed to fall from the fine lines of design. And accordingly words are marginalized in the surveyor's map in much the same way as decorative embellishment, usually through their sequestration in the field book.

Aaron Rathborne instructs his reader to number the 'closes' on his map, and to put the key to the tenants and tenancies in the field book. The number, he assures, 'will always direct you to your booke, where you may find it at large' (Rathborne 1616: 128). The latter 'ingrossed book' is the proper place for such information, for 'much writing in your plat' is 'absurd and grosse'; 'breedeth confusion, and causeth much cumber and trouble' (ibid.: 128). And such verbal 'ingrossement' registers the return of those legal judgements and negotiations that form the basis of traditional postfeudal stewardship. Ralph Agas, after comparing the map favourably to the 'auncient and faire bookes' used in manors, tells the surveyor to reproduce the map of his survey in a 'faire parchment booke with a large margent' (Agas 1596: 15). 'At every Alienation & change of Tenaunts, landeholders, and Farmours,' he explains, 'their names shal be registred and quoted in the same margent, concurring with their tenures' (ibid.: 15).

Self-fashioning and the middle man

The relationship between mathematical and verbal surveying; between map and field book, which persists throughout the seventeenth century, embodies the equivocal identity of the surveyor himself. Like the *'Gentleman-Tradesman'* in Defoe's *Moll Flanders* (1722), another marker of an equivocal modernity, the surveyor is a *'Land-water-thing'* (Defoe 1976: 60). He is both liberal scholar and wage labourer, and he is neither gentleman amateur nor common mechanic; neither landlord nor tenant. The value and meaning of his role, his work and its product depend on a precarious rhetoric of intermediacy, which figures the mathematician's art, much like the steward's, as somehow in between. And in an important sense, the surveyor remained a judicious steward, not just in the generic constructs of the surveying manual, but in the social fabric of the agrarian economy. The seventeenth-century surveyor, like the steward, was no free scholar or even professional, but remained tightly wedded to his patrons, their values and their land.

Of the two hundred surveyors identified by Sarah Bendall as working in Cambridgeshire between 1600 and 1836, about one-sixth can be defined more or less as professional mathematicians: schoolmasters, authors of surveying manuals and members of learned societies (Bendall 1996: 69). The rest 'supplemented their income from agricultural activities, as farmers or estate stewards, and many were occupied in other land-related activities' (ibid.: 69). Although surveyors exhibited increasing social and geographic mobility in the period; increased status; and greater freedom from patronage; this development was a gradual and principally eighteenth-century one. Most surveyors in the seventeenth century worked locally and for a tightly defined network of employers (P.D.A. Harvey 1996: 35). Despite their impressive cartographic record in a region of relative economic liberality, this attachment to local structures of patronage characterizes the work of the Walkers of Hanningfield. The Walkers worked almost exclusively after 1590 under the patronage of a circle of central and mid-south Essex landowners such as Sir John Petre, who between them 'dominated the Essex bench' (Edwards and Newton 1984: 20).

All estate maps, points out P.D.A. Harvey, have two 'parents': the surveyor and the commissioner of the survey, generally the landowner (P.D.A. Harvey 1996: 30). Exploring the diversity of seventeenth-century property inscription, from terriers of freehold farms to vast maps celebrating the moral economy of a manor, with all its Medieval customs, we can rarely be certain whose impulse determined the mode of representation. What we can be certain of is that this choice was the product of a negotiation between commissioner and surveyor (ibid.: 42–3). This negotiation, I think, is where mathematics assumed an importance far beyond its practical utility or indeed its actual deployment as an instrument of land management and reform. It is where the surveyor appealed to the vanity of his client by placing him on a panoptic chair. But it is also where the surveyor made clear the impotence of such panopticism without his own stewardly negotiations.

Surveyors like the Walkers made no bones about their stewardly status and their capacity to negotiate between the Landlord's geometric view and the social complexities of his estate. Indeed they made this negotiation conspicuous in their geographies. Rather than simply opening up manorial lands to unmediated view, the Walkers present their surveys to their patrons as acutely mediated texts, the products of arduous labour and negotiation: of painfulness and travail. They position themselves not as executors of neutral mathematical law, but as artful investigators and guides, both in the manor courts and in survey texts themselves. John Walker Junior writes:

> Forasmuch as it is a thinge usuall amongst manie Surveyours, and also in my Judgement most necessarie & needefull first to sett down some direcc[I]ons to be as it were a Lanthorne and light for the Lord or Owner of the landes of which any Survey is made.
>
> (Edwards and Newton 1984: 65)

A map, we might with some justification say, always has a margin, as surely, indeed, as a mind always has a body, and as the knowledgeable architect, painter or cartographer always has a crafty, dextrous hand. This margin, particularly generous in the early modern period, is the proper location for what positivist cartographic historians deem 'matter not forming part of the map itself' (Skelton 1952: 18). A crucial component of this marginal matter is the information and guidance sought from local inhabitants by the surveyor of an unfamiliar or socially complex domain. Upon this help he depended not just to make his map but even to find his way, where mathematics left him little better than transcendently lost.

Treatises in surveying advise the artist in mitigating the contingency of such guidance: he must, for instance, ask local authorities to choose for him the oldest and most honest inhabitants of a region (see Norden 1607: 24). If anything such attempts make still more glaringly apparent a rupture in the universal, self-sufficient authority of mathematics, into which seeps the discreteness of history and locale, and the stewardly discretion which must negotiate these vagaries. Historians following the lead of Brian Harley characteristically view such ruptures as evidence of the failure of cartographic discipline to suppress that resistance which it is designed to dominate and erase. My particular contention, against Harley's conception of the 'silent' disciplinary function of cartography, is that the early modern geographer makes no attempt to hide his precarious negotiation between local stewardship and liberal science. In fact it is crucial to his self-fashioning.

John Norden's master-excuse for a range of the faults and inconsistencies people have discovered in the first maps of his *Speculum Britanniae* is that he has had to depend in his surveying on guidance from such local people:

as by discretion of men in Aucthoritie are thought fit to yeelde me direct information, who yet thru their simplicitie or partialitie, may miscarrie the most provident observer, holding that to bee in their conceites of moment, and of the contrarie, as their affections leade them.

(Norden 1596: 17–18)

Norden's apology reveals not just the social politics inextricable from what was still only an embryonically 'modern' scientific exercise, but also the inextricably social and particular; the stewardly nature of the science to come. Almost as often as early modern cartographers boasted about the capacity of the map to elide the vicissitudes of actual travel, they complained noisily of the impossibility of making maps without the help and advice of reliable local inhabitants. Far from a lone admission of difficulty and failure, Norden's apology is typical of the seventeenth-century mathematician's promotion of himself as middle man: holder of the 'lanthorn' for his customers and employers.

The doubtful traveller

As in England, so in America. Whilst it would be facile to suggest that the American geographer was the exact social equivalent of his English counterpart, we can at least be certain that he operated according to analogous systems of patronage, and was similarly obliged to negotiate his own role and the nature of his work with a commissioning client. Where an American surveying profession finally and patchily evolved, in the latter decades of the seventeenth century, it tended, at least at first, to exhibit the same old-fashioned characteristics of versatility, amateurism and adherence to local networks of authority that we have remarked in England (Buisseret 1996b). William Goodsoe, town surveyor of Kittery in Maine from 1694 to 1715, seems to have developed his fairly primitive and highly versatile surveying work as an adjunct to a colourful career in the local court. Goodsoe appears in the records before his appointment both as a proven burglar and absentee from worship, and in a variety of less disreputable roles including those of surveyor and 'self-styled courtroom lawyer' (Candee 1982: 19). His mathematics secured him a relative authority in the legal processes of local government, most remarkable perhaps for its transcendence of Goodsoe's dubious past. Nonetheless this authority depended on the local legal context of the Kittery court itself, beyond which Goodsoe's primitive maps are and would have been largely illegible.[9]

Like their English counterparts, American geographers tended to be court men rather than liberal scholars or anonymous mechanics, their authority constituted in localized networks of influence. Like them too, they were ultimately rather disinclined to surrender this localized authority to mathematics. Far from trying to reduce and transcend the particularities of their context in the interests of mathematical authority, surveyors operating in

America were often keen to emphasize precisely the tricky particularities of their work.

John Love worked as a surveyor in North Carolina and Jamaica, and subsequently returned to England to write *Geodaesia*, a manual designed specifically for America (Benes 1981: 102). The relationship between geometry and America in this treatise seems reciprocal. The colonist needs geometry: 'How,' asks Love, 'could Men set down to Plant, without knowing some Distinction and Bounds of their Land?' (Love 1688: n.p.). But if the colonist needs geometry, the geometrician also needs America, as the blank sheet for his abstract designs. Indeed Love, concluding a surveying treatise in which America has allowed him space to explore the full range of planimetric possibilities, assures his reader complacently: 'so will you have a fair Map … better done, I think, than in any place of the World yet, except for the Harbours of *Eutopia*' (Love 1688: 194). Yet Love added an appendix to the second edition of his book advocating a cruder, more pragmatic form of surveying than the utopian geometries his first edition had projected. Love advises his reader that he may as well survey by the compass and chain alone, or even with 'a Pole or Rod cut out of the Hedge':

> You may have, I suppose, in *Crooked-Lane*, a Rod made to shoot one Part into another like a Fishing-rod, to be used as a Cane, in the Head whereof may be a small Compass. Which alone is Instrument enough to survey any Piece of this Earth, be it a Mannor or larger. And if so, what need is there of a Horseload of Brass Circles, and Semicircles, heavy Ball-sockets, Wooden Tables and Frames, and 3 legged Staffs, *cum multis aliis*, unless to amuse the ignorant Countryman, to make him more freely pay the Suveyor.
>
> (*Appendix*, Love 1715: 7)

Love's pragmatic revision of the American surveyor's work is confirmed in a poem inscribed upon 'an anonymous 1730 poem on a map of "25 Divisions" in Stoughton, Massachusetts':

> Upon our NEEDLE we depend
> In ye THICK WOODS our COURSE to know
> Then after it yᵉ CHAIN Extend
> For we must gain our DISTANCE so
>
> After yᵉ HILLS through BRUSHEY PLAINS
> And HIDEOUS SWAMPS where is no TRACK
> Cross RIVERS, BROOKS we with much PAINS
> Are fore'd [*sic*] to travel forth and back.
>
> …
>
> When WEARY STEPS has brought us home
> And NEEDLE, CHAIN have some respite
> SCALE and DIVIDERS in use come
> To FIT all for next morning light

And though we're CAREFUL in ye same
As HAST & OBSTICLES will yield
Yet after times they will us BLAME
When ROUGH WILD WOODS are made a Field.
(quoted in Candee 1982: 11–12)

Like the equivocal points and lines of seventeenth-century mathematical writers from Ramus to Dee I think pragmatic, dirty geometries such as these were intended to thicken out the cartographic text, reminding the commissioners of colonial maps that their view from the panoptic chair was impotent without the mediating agency of the tough, experienced surveyor.

Produced before the home-grown surveying tradition had become widespread, those rare and exceptional public maps on which our understanding of Anglo-American geography is based are often still best understood in terms of the delicate self-fashioning of the middle man.

Throughout the seventeenth century, both cartographers and those who commissioned them sought to gain leverage through their maps with powerful patrons and institutions.[10] John Smith's geography is the product of a relatively liberal, humanist conception of the potential of America to accommodate English settlers alongside reformed American natives. But it is simultaneously a concerted piece of self-fashioning. In one respect, it is precisely this attempt at self-fashioning which has led to the notoriety of Smith's geography as a work of ideological erasure. Before handing the map he titled *New England Observed and Described* over to its Dutch engraver in 1616, John Smith presented it to Charles, Prince of Wales, offering him the opportunity to change its native names to English, 'that posterity may say, King Charles was their godfather' (quoted in Benes 1981: 5). In general, however, Smith's self-fashioning worked, alongside his humanist optimism and mercantilist conception of colonization, not to erase Indian geography but to reinforce it.

Smith's writing is notorious for foregrounding its author's own role in preserving the early Jamestown colony, intensifying this emphasis in later editions of early tracts. Presenting himself as the exemplum of the successful colonist, Smith seeks both to influence the future direction of American settlement, and to raise royalties and sponsorship for his own part in it (Hayes 1991; Canny 1988: 214). The qualities Smith associates with successful settlement and exemplifies in writing up his own exploits are diligence, ingenuity and justice, not least in handling relations with the Indians. Smith expresses impatience and contempt in his various writings for those superiors amongst the colonists who refuse to work; lust after gold; and – in the case of Jamestown governor Captain Newport – insist upon following orders from London to the letter with no reference to the specific conditions in and character of Virginia and its native inhabitants (Hayes 1991: 142–3). Foregrounding Indian geography, Indian customs and his own encounters with Indians, Smith fashions himself as the opposite of Newport: one who can adapt the requirements of his sponsors to conditions from which they are utterly divorced.

Like the English surveyor, Smith wishes to place the audience for his American geographies on a geometric 'chair' from which they have a clear view of their possessions. But he also wants to remind them how removed this position is from any effectual agency in the American landscape itself. In *The Generall Historie of Virginia, New England, and the Summer Isles* which Smith published in 1624, casting a retrospective eye back over his experiences, Smith makes clear that he himself is 'no scholar', but one who occupies, in Kevin Hayes's words, a 'kind of middle ground' between sterile refinement and uneducated craft (*The Generall Historie of Virginia, New England, and the Summer Isles*, Smith 1986, II: 188; Hayes 1991: 142).

We might read the large compasses figured on Smith's maps of Virginia and New England as another such reminder. In *Hakluytus Posthumus* Samuel Purchas gives the conventional gloss for this common poetic and cartographic trope (Figure 7.1): 'and as in Geometricall compasses one foote is fixed in the Centre, whiles the other mooveth in the Circumference, so it is with Purchas and his Pilgrimes, in this Geographicall compassing' (Purchas 1905–07, XX:

Figure 7.1 Geffrey Whitney: detail of title page in *A Choice of Emblemes and other devises for the moste parte gathered out of sundrie writers* (Leyden, 1586). Reproduction courtesy of the British Library (shelfmark: 89.e.11).

130). Smith presents himself in his writing not simply as the still foot – the Lord on his chair, seeing through geometry – but also, and most importantly, as the moving one, covering real territory and learning through experience. And the successful colonist and geographer cannot rely on sterile scholarship alone, any more than he can rely blindly on the instructions of superiors in London, but must be prepared occasionally to leave his compasses at home.

Kevin Hayes has observed that Smith gives three accounts in his various re-writings of an episode in which he was captured by the Indian Opechan-canough, from whom he subsequently escaped (Hayes 1991). In the early versions of the story Smith tells of dazzling his captor with a compass; in the final version he leaves out the compass altogether, attributing his escape not to the 'magic' of science, but to personal qualities of ingenuity, discretion and experience (ibid.: 133–5). In this last, perfected version of his captivity Smith 'speaks the language' of the Indians he negotiates with. Setting aside his compasses he is able to work with the dirtier geometries of Indian infor-mants such as the chief Powhatan, who during Smith's captivity 'began to draw plots upon the ground' (*The Generall Historie of Virginia, New England, and the Summer Isles*, Smith 1986, II: 183).

Recent scholarship has been much preoccupied with these native maps which it takes largely to be obscured and parasitized by colonial geography, viewing mentions such as Smith's as rare and guilty ruptures in a prevailing wall of silence. I think no such false consciousness was involved. Setting aside his compasses, Smith deliberately embraced a dirty native geometry which complemented his own unscholarly, pragmatic bent. Verses added to a description of his map celebrate this unique access into an un-geometrized, un-European world: 'Thus have I walkt a wayless way, with uncouth pace, / Which yet no Christian man did ever trace' (*The Generall Historie of Virginia, New England, and the Summer Isles*, Smith 1986, II: 107).

Contrary to the dominant, Harleyan view I think that it was not simply through some transcendent, geometric overview that the geographer assumed agency in America, but also, and more characteristically through getting lost: through setting their compasses rhetorically aside. William Wood's *New England's Prospect* (1634) is another geography conspicuously prepared to step down from the panoptic study chair. Wood judges Indian guides, recall, to be as well acquainted with their environment as the 'experienced citizen' with Cheapside. He also tells a story which confirms his preparedness to abandon science and trust Indian geography. 'The doubtful traveller', he records, 'hath oftentimes beene much beholding to ... [the Indians] for their guidance thorow the unbeaten wildernesse' (Wood 1634: 70). 'My selfe,' he continues, illustrating his point:

> with two more of my associates bending our course to new *Plimouth*, lost our way, being deluded by a misleading path which we still followed, being as we thought too broad for an *Indian* path ... which begat in us a security of our wrong way to be right, when indeed there was nothing

lesse: the day being gloomy and our compasses at home ... Happily we arrived at an *Indian Wigwamme* ... The son of my naked host ... took the clew of his travelling experience, conducting us through the strange labyrinth, of unbeaten bushy wayes.

(Wood 1634: 70)

Like John Smith, Wood constructs his discrete and exclusive authority as an American geographer not simply in having a set of compasses, but in leaving them at home. He fashions himself more as a stewardly intermediary between his educated, distant readership and the local, customary knowledge of his 'naked hosts', than as surveyor of some geometric *tabula rasa*.

Dangerous geographies

Sometimes the American geographer's mediation was far from benign: a geography of violent conflict rather than of lost-ness and being guided. But I think the dangerous geographies of colonial warfare can still be read as part of a careful strategy of self-fashioning, distinguishing discrete judgement and experience from the impotent geometries of armchair travellers.

'A specific correlation exists,' wites Peter Benes, 'between New England mapping and the expression or resolution of power conflicts' (Benes 1981: xviii). In the 1630s the embryonic New England colonies forged their earliest self-definitions through a strange interplay between geography, narrative and violence.

In 1637 around 400 Pequot Indians burned to death in their stockaded village on the site of the present-day town of Mystic, Connecticut. This catastrophic event represented the pinnacle of a crisis that had been building for some time. Struggling with their tribal neighbours to control the beaver trade with the Dutch, posted since 1614 on the Connecticut river, and losing patience with the encroachments of English settlement, the Pequot had launched a series of small-scale offensives that led to a confederation of New England colonies and local Indian tribes against them; to the 1637 so-called 'Pequot War'; and ultimately to the Mystic massacre.

According to contemporary accounts, at dawn on the morning of 26 May Connecticut troops led by Captain John Mason, and Massachusetts troops led by Captain John Underhill, together with Indian allies from the Mohegan and Narragansett tribes, surrounded an Indian stockade just above the Mystic river. They took the fort by surprise, Mason entering from one side, Underhill from the other, and a ring of English soldiers and Indians remaining outside. It was Mason that took the decision to set fire to 80 huts within the fort, housing men, women and children. Most died within an hour. And most of the few that escaped were killed by the Indians and English soldiers surrounding the fort.

The first accounts of the Mystic massacre were written immediately after the event by Philip Vincent, an English visitor to New England, and by Captain John Underhill, leader of the Massachusetts troops. Both were published in

London in 1638. Eighteen years later, Captain John Mason of Connecticut wrote his own history of the Pequot War – to correct, he said, the 'random' nature of previous accounts (implicitly Underhill's and Vincent's). Mason's history was ultimately printed as part of another text in 1677 (Hubbard 1677). In it he records with chilling impersonality his judgement on the Pequot: 'The Captain also said "We must burn them"' (Mason 1897: 28).

Random or not, Underhill's and Vincent's contemporary accounts of the Pequot War have been treated as something of an anomaly by modern studies of early American history and culture. The anomaly is comprised in their uncertainty between history and geography. *Regeneration Through Violence* (1973), Richard Slotkin's seminal study of New England history and the Puritan mind, talks about the way discourses on 'land' 'disrupt' Vincent's war narrative (Slotkin 1973: 69). Wayne Franklin's *Discoverers, Explorers, Settlers* (1979) describes Underhill's narrative as 'split between martial and peaceful impulses, promising as it does on its title page 'both a "true narration" of the war and "a new and experimentall disoverie"' of previously unexplored terrain (Franklin 1979: 43). Franklin countenances Underhill's excuse that a lack of time has forced his failure to separate description from narration – at least into two separate sections of his book. But he also thinks that the relationship between history and geography in Underhill's text is more integral than this excuse suggests. In Franklin's words, 'Underhill finds in the snatch of a future horizon the necessary oblivion for a historical struggle he would rather forget'.

Franklin's Underhill seeks escape from history into geography as a means of forgetting the trauma of war, and mapping out the prospect of a less troubled colonial future. Geography allows him to imagine the clean, disciplined 'tabula rasa' which Slotkin takes to be the precondition for colonial fantasy (Slotkin 1973: 38). But I think Underhill is deliberately presenting us with a far darker, dirtier vision than this: one in which narrative history, geography and violence remain intimately entangled, and in which geography is neither 'peaceful' nor 'oblivious'.

We can read Underhill's history of the Pequot War as a simultaneously geographic text in various ways. Most obviously, Underhill's stated purpose has been to 'interweave' a 'true relation' of the violent events of 1637 with descriptions of 'special places fit for new plantations': clean prospects of the future (Underhill 1897: 49). But as well as alternating his history and geography, Underhill's history itself has a distinctly geographic emphasis. Underhill documents the journey of the forces allied against the Pequots through territory unfamiliar to European eyes. He even tells the tale of two Puritan 'maids' captured by and ransomed from the Pequots as a story of geographic discovery: 'The Indians carried them from place to place, and showed them their forts and curious wigwams and houses' (Underhill 1897: 72). And the kernel of Underhill's journey and narrative of discovery – Mystic fort itself – is represented in a kind of map (Figure 7.2).[11]

Underhill's 'map' of Mystic fort might seem to us much like Mason's objectification of his own brutal words and deeds ('the Captain also said ...').

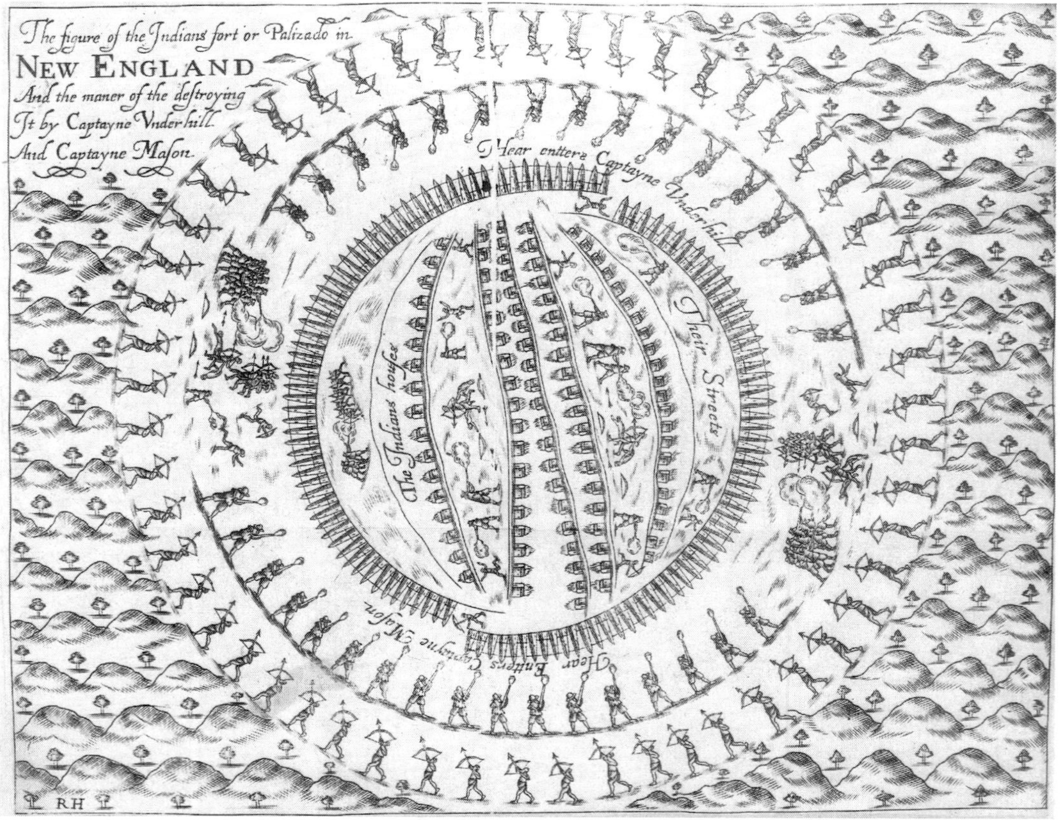
The figure of the Indians fort or Palizado in
NEW ENGLAND
And the maner of the destroying
It by Captayne Vnderhill
And Captayne Mason
Hear entters Captayne Vnderhill
Their Streets
The Indians houses
Hear Entters Captayne Mason
RH

Figure 7.2 John Underhill, 'The figure of the Indians' fort or Palizado in New England . . .', in *Newes from America; or a new and experimentall discoverie of New England* . . . (London, 1638). Reproduction courtesy of the British Library (shelfmark: c.33.c.25).

It's an attempt to gain separation from and perspective on a haunting, claustrophobic experience and an advertisement of the panoptic power to gain this separation: to map the dark heart of the American wilderness. This advertising function conditions its relative sophistication. Peter Benes finds the geometric style of this representation, 'drawn for an English audience by an English engraver', markedly distinct from the functional crudity of most vernacular American maps (Benes 1981: xviii). But whilst Underhill's narrative and map adopt the distanced, public perspective of the geographic mode, they don't forget history.

Underhill's map of Mystic is a representation not just of place but also of historical event. And if it doesn't de-temporalize its object, as modern geography is supposed to do, neither does this map de-localize or de-socialize it. The Pequot inhabitants of Mystic fort, and the violence done to them, are very much at the centre of Underhill's map. They are what define and give identity to this place and to the Puritan engagement with it. This is still more explicit in Vincent's narrative, which almost lovingly fills in the lines which Underhill's map leaves relatively empty. Vincent writes:

> Let me now describe this military fortress, which natural reason and experience hath taught them to erect, without mathematical skill, or use of iron tool. They choose a piece of ground, dry and of best advantage, forty or fifty foot square (but this was at least two acres of ground.) Here they pitch, close together as they can, young trees and half trees, as thick as a man's thigh or the calf of his leg. . . . The space therein is full of wigwams, wherein their wives and children live with them. These huts or little houses are framed like our garden arbors, something more round, very strong and handsome, covered with close-wrought mats, made by their women, of flags, rushes, and hempen threads, so defensive that neither rain, though never so strong, can enter. The top through a square hole giveth passage to the smoke, which in rainy weather is covered with a pluver. This fort was so crowded with these numerous dwellings, that the English wanted foot-room to grapple with their adversaries, and therefore set fire on all.
>
> (Vincent 1897: 105–6)

Vincent's description of Mystic explicitly confirms both the local particularity and the absolute humanity of a social space whose annihilation his narrative will also shortly reproduce. This geographic scene does not and cannot shed the burden of its past: an inextricably Indian past of use and habitation of the land. Moreover Vincent's Indians do not just comprise the 'content' of this scene: they parasitize its frame. Vincent is able, like Underhill, to take the measure of; to rationalize Mystic fort because Mystic fort is already a rational construct. The 'natural reason and experience' of the Pequots, expressed in the framework of their settlement, means that his geographic perspective is also theirs.

I don't think the geography in Underhill's bifurcated text is as clean; as silent; as peaceful or forgetful as either Franklin or Slotkin suggest. It seems to me still much inhabited both by Indian history and by colonial violence. I also think that Underhill's vacillation between history and geography, rather than a disruption in an unfinished text, is a fairly typical characteristic of early American representation. When Mason's narrative eventually appeared in print, just under 30 years after Underhill's and Vincent's, it was as a supplement to the Massachusetts Puritan William Hubbard's 1677 account of the Indian troubles of his own age (Hubbard 1677).[12] Hubbard also supplemented his principally narrative text with a woodcut 'map of New England', made by one John Foster, and – its cartouche proudly claims – 'the first that was here cut'.[13]

Hubbard's map registers not just the Narragansett Indian presence in the region – the source of the present troubles – but also Pequot territory, when this tribe could already be described in 1638, in Vincent's narrative, as 'nothing but a name' (Vincent 1897: 107). Naturally we can see in practical terms why the Pequot feature on Hubbard's map. They are present to 'set the scene' for the readers of Mason's narrative, which in its turn sets the scene for Hubbard's. But one of the effects of this inclusion is to mark the land again with a history which is inextricably the history of the Indian tribe and of the violence done to them. And far from using this map to forget the trauma of this kind of history, or to eradicate its contradictions, Hubbard works diligently to fill it with more history than the cartographic frame can conventionally contain. In effect, he supplements his supplement.

Hubbard's map, marked with over 50 numbered locations, is attended by a table which decodes these encryptions. Each entry is a dense account of some event in the Puritan/Indian conflict of the 1670s, often making reference not just to locations in the map, but also, for more extensive treatment of the event, to locations in Hubbard's main narrative. Some of these entries are so resonant with the particularity of time and place as to be moving. One tells of a woman captured by Narragansett Indians and asked to milk cows stolen from her neighbours (Hubbard 1677: 133). Another reads:

> 4. *Dartmouth*, where in *June* 1676, a man and a woman were slain by the *Indians*; another woman was wounded and taken; but because she had kept an *Indians* child before, so much kindness was shewed her, as that she was sent back, after they had dressed her wound; the *Indians* guarded her till she came within sight of the English.
>
> (Hubbard 1677: 132)

Hubbard's table of events seems to want to overcome the difference; to build a bridge between geography (his map) and history (his main narrative). And as it builds this bridge it also discovers the relation between the Puritan colonists of New England and the physical and social space that they inhabit. This relation is not the clean separation of the geometric line, transcendently framing and forgetting its particular content. It is, rather, a relation of profound particularity and intimacy. Predominantly this intimacy is the awful intimacy of violence. But it is also, occasionally, the intimacy of cooperation and exchange.

Matthew Edney and Susan Cimburek have studied Hubbard's map as both 'a cultural construction of regional identity' and 'an integral element of political and religious discourse in later-seventeenth-century New England' (Edney and Cimburek 2004). Hubbard's map, they argue, contributed to his 'construction of a particular identity for the English colonists in New England as religious martyrs' (Edney and Cimburek 2004: 325). Its geographic digressions into historical event reflect the same Puritan preoccupation with Providential chronology that structures Edward Johnson's chorography of New England. Yet as in Johnson's account these reductions are far from total. Whilst certainly providentialist, Hubbard, like Edward Johnson, is not inclined to reduce New England history and geography to the spiritousness of pure typology, tending instead to celebrate the material successes of his region (Edney and Cimburek 2004: 343). Moreover, at the same time as he attempts a characteristic Puritan balance between improvement and reduction, Hubbard treads a familiar geographic tightrope between scholarly authority and stewardly discretion.

Hubbard's cartographic geometry, suggest Edney and Cimburek, is part of a conventional 'self-effacing rhetoric' which allows him to enter public discourse, speaking a truth which can be verified by an international constituency of educated peers (Edney and Cimburek 2004: 342). Yet Hubbard also limits the authority of his geometry, making excuses like so many other geographers which draw attention to dangers and difficulties which few metropolitan mathematicians would be prepared to face. If he has committed any error, Hubbard reasons in his 'Advertisement to the Reader':

> about the Scituation or distance of places, it may deserve an excuse rather than a censure: For our Souldiers in the pursuit of their enemies being drawn to may desert places, inaccessible Woods, and unknown Paths, which no Geographers hand ever measured, scarce any vultures eye had ever seen, there was a necessity to take up many things in reference thereunto upon no better credit sometimes then common Report.
>
> (Hubbard 1677: sig.A2r)

William Hubbard's geography reminds his English and American readership that they are all 'doubtful travellers' in a wilderness which God's providence is only gradually illuminating. It also reminds them that the thin lines of imperial geometry are often impotent in such places, and that the metropolitan empire needs its diligent and canny lanthorn-bearers. In its Daedalean rhetoric of mediation between space and time, between scholarship and experience, between clean virtue and dirty profit, it is a fitting case with which to end this history of the doubtful, sandy, poetic geometries of the English seventeenth century.

Notes

1 Introduction

1 Matthew Edney has compared the positivist history of cartography to the Whiggish history of constitutional liberty in its celebration of the irresistible shedding of shackles of prejudice and ignorance (Edney 1993: 57). He argues that it conceals a reality which is uneven and heterogeneous, treating maps as part of a unified Enlightenment project of assembling geographic data, and regarding such data as if it existed in its own world, like 'Platonic forms', detached from the interests and material circumstances of individual instances of cartography (Edney 1993: 55). After 1800, he notes, maps began to be catalogued consistently according to the region they dealt with rather than the genre, project or period they originated from, as if one map was very much like another (Edney 1993: 64).

2 Where Bernhard Klein remarks, in *Maps and the Writing of Space* (2001), that his discussion of cartography 'is ultimately less motivated by an interest in the history of *maps* than in the culture of mapping', he is clearly making a commitment to a more generalizing mode of history than, for instance, Jerry Brotton's (Klein 2001: 8; Brotton 1997). In Klein's work, as in the work of Richard Helgerson, John Gillies, Tom Conley, Garrett Sullivan and Rhonda Lemke Sanford, all literary scholars of cartography, subscription to this notion of a general 'culture of mapping' is part of a broader willingness to consider history in terms of a 'poetics' of culture (Helgerson 1992; Gillies 1994; Conley 1996; Sullivan 1998; Sanford 2002). Such work is likely to go looking for, and is likely to find, what Helgerson calls a 'deep, if unexpected likeness' between categories of texts held separate by disciplinary boundaries, and especially between literature and geography (Helgerson 1998: 14).

3 Matthew Edney has also found Harley falling into the trap of an alternative positivism in his attempt to compensate for the malign, or 'silent' cartographies of the past (Edney 1996: 188).

4 The structuralist tradition of literary analysis establishes particular relationships between individual literary texts as intersections within wider differential systems of cultural communication. Whilst this structuralist 'reduction' of history in literary studies is now largely a thing of the past, it has developed into a sophisticated dialectical account of the cultural uses of literary form which has revealing things to say about geography. The literary historian Ian Watt some time ago identified a new emphasis on recognizable, geographic, rather than fanciful, romance setting in prose fiction from the early eighteenth century onward with the formulation of a new, novel genre constituted through a matrix of this and other elements of 'formal realism' (Ian Watt 1957). In turn, Watt relates these generic similarities in new prose narratives to contemporary social and philosophical developments: the rise of a Protestant middle class demanding new stories to be told than myths of aristocratic virtue, and the concurrent rise of a Lockean empiricism which judged truth and character to be

constituted and discovered through human experience and the facts of individual circumstance; not fixed and self-evident in the class stereotypes of romance. Michael McKeon has described the relationship between genre and history in similarly dialectical terms, as both product and representation of shifting notions of 'truth' and 'virtue' (McKeon 1987). For McKeon new, empirical notions of truth, and new, middle-class Protestant notions of virtue called for new stories to be told, and hence the new genre of the novel was born. In turn, the new genre shaped the way human stories could be told, and perpetuated such truths and virtues. For McKeon genre is itself a map of cultural meaning, rather than a container for such cultural maps, although it also has a certain life of its own: a persistence beyond the immediate bounds of local cultural determination.

2 Discipline and polish

1 Whilst earlier governmental responses to enclosure had sought to suppress a phenomenon perceived as detrimental to the national interest, and to this end set up commissions of inquiry in 1607, 1630, 1632, 1635 and 1636 to look into depopulating enclosures, the last parliamentary proposals to deal with enclosure as a 'problem' were rejected in the 1650s as a threat to 'property', and when, in 1660, an act was passed abolishing feudal tenures, it did not safeguard the rights of copyholders, leaving them subject to eviction (Darby 1973: 21; Christopher Hill 1980: 13, 21, 127). By the last few decades of the century the tide had definitively turned: the new government of landlords ensconced at the Restoration passed no anti-enclosure legislation, and the Hanoverian era saw the majority of enclosures licensed by Parliamentary statute (Christopher Hill 1980: 128).

2 See Slotkin 1973: 48–9; Jennings 1975: 71–6; Hulme 1981: 69; Jehlen 1993: 41–2; Mackenthun 1997: 43; McLeod 1999: 11; Seed 2000: 206.

3 Hartlib himself worked closely in the 1630s with mathematical practitioners and inventors (Webster 1975: 358). Between 1630 and 1634 his correspondent John Pell conceived a plan, laid out in 'An Idea of Mathematics' (1650), for a 'mathematical intelligence centre' which would collect books and instruments, compile reports, maintain correspondence with leading mathematicians, act as an 'employment exchange' for mathematical practitioners, and exert quality control on the profession (Webster 1975: 356–8).

4 Patrick Collinson has suggested that the signature cultural practice of Puritanism is fasting: the reduction or refinement of the body to render it less worldly and more fit a receptacle for divine grace (Collinson 1996). Puritan strictures on dress and hair style represented further pressures to reduce the excrescent, worldly elements of the body. In these austere practices, and in the strict accountancy which they practised upon themselves in attempts to perceive the signs of salvation or damnation, Puritans worked to reduce the corruption of their own bodies and lives to disciplined and coherent form. In their typologies they worked to reduce the chaos of personal and public history to the good order of scriptural parallel and eschatology. Directing their reductive perspective towards the institutions and practices of church and state, Puritans campaigned unsuccessfully from Elizabeth I onward to reduce church liturgy, and indeed all moral authority to the unembellished sanction of scripture, just as they strove to trim churches of their extraneous decorations. They became notorious for the wider antipathy this reductionism licensed to any customary authority or practice, from the unscriptural rites of birth, marriage and death, to the maypoles and Whitsun ales of rural folk tradition. Millenarianism, writes Charles Webster, 'generated a more critical and aggressive attitude towards all agencies which were recognized as obstacles of the Reformation' (Webster 1975: 7).

5 Perry Miller's scholarship on New England Puritanism largely takes at its word the seventeenth-century 'declension' thesis (Perry Miller 1956).

3 Humanist geometries

1 See Hooykaas 1958 for a thorough account of Ramus's mathematical works.

2 Any translations from Hooykaas are my own.

3 In footnotes to a recent essay Cormack draws up sides between the Yatesian history which encouraged us 'to see Dee as an Elizabethan "magus"', and Sherman's more 'balanced' account which has 'set the record straight' by 'establishing the primacy' both of Dee's 'geographical work' and of his social relationships with 'court and government' (Cormack 2001: 65). Cormack's assurance about the 'true', pragmatic nature of Dee's mathematics does not utterly disperse her evident perplexity that though he 'seems to have kept separate' magic and geography 'in his dealings with geographers,' the symbolism of his texts 'argues for a closer relationship' (Cormack 2001: 66). Moreover, this uncertainty about Dee's geography extends to Cormack's account of geography in general. Despite arguing the artificiality of any separation between scientific theory and practice and avowing the primacy of utility in early modern geography, Cormack also defines geography as 'an intermediate discipline that combined aspects of theory and practice': one whose 'ideology', as she puts it, expressed in paratexts such as Dee's own taxonomy of mathematical science, placed 'an immense value' on mathematics (Cormack 1991: 640).

4 In *De Corpore*, published in 1655, Hobbes argued that since only body was real, mathematics must be the generalized science of body (Jesseph 1999: 70). For Hobbes, as for Ramus, the Aristotelian and scholastic distinction between pure and applied mathematics was meaningless: all mathematics was measurement, whether applied to physical or imaginary objects (Jesseph 1999: 74, 135). Since there was no higher human faculty than the imagination – no intellectual faculty, that is – there was nothing for 'pure' mathematics to study. Moreover, as with Ramus, once again, and as with Bacon's humanist science, Hobbes believed a mathematics derived from human use must be returned to human use: applied to the end of gaining power over nature (Jesseph 1999: 191). This materialist mathematics required a reappraisal of the implicitly idealist principles of Euclidean geometry. Hobbes redefined the Euclidean indivisible point as a body whose magnitude is set aside, and the indivisible line as engendered by the movement of the point (Jesseph 1999: 76–7).

5 Charles Webster describes a comparable debate between the Hartlib circle and the Cambridge Platonist Henry More, who referred to Baconian experimenters as men who 'dig and droyle like blinde moldewarpes in the earth' (quoted in Webster 1975: 147).

6 David Armitage argues that the category of 'literature' is projected onto early modern writing as anachronistically as the category of 'empire' (Armitage 1998).

7 The irony may be unintended, since Hobbes's mathematical materialism was only fully outlined in *De Corpore*, published three years after *Oceana*. On the other hand, Harrington may be taking Wallis's line, which viewed material points and lines as indeed a geometry of nothing.

8 See P.D.A. Harvey 1980 for the history of surveying in Egypt.

9 See Orgel 1965: 151 for the classical provenance and typical Renaissance interpretation of this theme.

10 In his patronage of John Speed Greville might be thought to have facilitated a literal version of such a humanist geographic overview.

4 Discipline reconsidered

1 See, for instance, 'A Rapture', Carew 1949: ll. 17–20, p. 49; 'Britannia's Pastorals', Browne 1868–69, I: book I, song 4, p. 101; Manlius and Sherburne 1675: 5.

2 Sullivan devotes a chapter to this play (Sullivan 1998: 159–93).

5 Ambivalent geographies

1 Cecil spoke in Parliament in 1601 to support legislation designed to protect 'tillage' against the incursions of enclosure (McRae 1996: 8–9).

2 In this conservative sentiment Norden is in agreement with the surveyor Ralph Agas, who wrote in 1606 'against the turning of copyholds into freeholds' (Agas 1606).

3 'The dialectic of general and particular that is built into the structure of a chorography,' concludes Helgerson, 'in the end constitutes the nation it represents' (Helgerson 1992: 138).

4 Sarah Tyacke's *English Map-Making 1500–1650* (1983) places a detail of a Walker map alongside William Leybourn's generic model of an estate map, whose empty outlines the Walker map appears to fill in (Tyacke 1983: Figure 10, Figure 11). Whilst there are strong resemblances, and the Walker map clearly stands, as the caption notes, as 'an emblem of pride in property and ownership of the land', it is also clear that this pride is constituted in the preservation and foregrounding, rather than erasure, of the customary integrity of the manor.

5 See also the 'Title panel from Samuel Walker's map of the manor of Garnetts, Essex, 1622' for a similar declaration (Tyacke 1983: Figure 4).

6 David Buisseret warns that estate maps, in particular, are impossible to understand divorced from 'map' and 'social' contexts (Buisseret 1996a: 1).

7 The period between the Tudor enclosures, spearheaded by much-vilified individual lords, and the sweeping parliamentary enclosures of the eighteenth century was chiefly that of 'enclosure by agreement': the process whereby freeholders contracted to extinguish common rights and re-allocate their holdings as consolidated enclosures, under the supervision of a commission of arbitrators whose decisions were ratified by courts of Chancery or Exchequer (Butlin 1979: 66, 68, 76; Christopher Hill 1980: 37).

8 See Corcoran 1992 for an excellent account of Petty's Down survey, and of the lessons Thomas Holme derived from it for his work in Pennsylvania.

9 See Fairbanks and Trent 1982: Figure 19, pp. 31–2 and Benes 1981: Figure 33, p. 36 for reproductions of and notes on the Chelmsford plan.

10 Hildegard Binder Johnson notes that 'landholdings were assigned and measured by the metes-and-bounds system' throughout the colonies (Johnson 1976: 25–6).

11 See Allen 1982: Figure 28, pp. 38–9 for an example of a dispute provoked by poor original surveying.

12 See Allen 1982: Figure 24, pp. 34–5 for a 1682 example of the more rigorous surveying practice called for by Massachusetts General Court at the end of the seventeenth century.

13 See Wallis 1985: 39–41 for the role of the Hakluyt cousins in promoting Raleigh's voyages.

14 Bruce McLeod reads Harriot's frontispiece, with its ornamental Indians and architectural motifs, as exemplary of an attempt at 'bringing order to confusion … the taming of nature, the transformation of perspective whereby the imperial English self is left in control of space' (McLeod 1999: 65).

15 For accounts of the surveying work which produced Smith's map and of its wide circulation and 60-year lifespan as the foremost map of the region, see Penrose 1952: 240; Verner 1968; Cumming 1972: 17. For a reproduction and notes see Morrison *et al.* 1983: Figure 7, p. 11.

16 Smith's map was preceded by a manuscript map of the Chesapeake Bay and Virginia made between 1608 and 1610 by George Percy, governor of Virginia in 1609–10, probably for his successor, Sir Thomas Gates (Morrison *et al.* 1983: Figure 6, p. 8). Gates's map strives, like Smith's, for a detailed Indian geography.

17 Harley notes similar 'admissions' from contemporary Spanish and French cartographers of North America (Harley 2001: 173–4). See Waselkov 1998: 207–12 for an

account of Smith's geographic relations with Indians; the Indian contributions about which he was so 'exceptionally forthright'; and Smith's iconographic representation of these relations in his map. See also Rountree and Turner 2002 for an assessment of Smith as a reasonably fair reporter of Powhatan Indian culture.

18 Bruce McLeod, who traces an 'apparent contradiction' in various early accounts which both acknowledge and discount Indian geography, finds this contradiction 'resolved because of Indian violence' (McLeod 1999: 94). 'By the actions of 1622,' writes McLeod, '*all* Indians had removed themselves from *any* juridical system' (McLeod 1999: 94).

19 See Benes 1981: Figure 5, p. 8 for a reproduction of and notes on William Wood's woodcut map *The South part of New-England, as it is planted this yeare, 1634*.

20 William Strachey, First Secretary to the Colony of Virginia, begins *The Historie of Travell into Virginia Britania* (1612) like so many other promoters with an acknowledgement of the general 'Clamour' against the '*vnnationall, and vnlawfull*' nature of the English '*Clayme*', perceived by many as 'a Trauayle of flat Impiety, and displeasing before god' (Strachey 1953: 7, 9).

6 Points mean prizes

1 Lisa Jardine describes the two-tier educational system operated by Dr Richard Busby, Master of Westminster School from 1638–95: 'a standard classical education for the well-to-do' and a 'technical, mathematical training' for boys such as Robert Hooke, selected by Busby for their talent, and taught with texts including Billingsley's Euclid (Jardine 2003: 61). Later, in the Royal Mathematical School established within Christ's Hospital School in 1673 by Samuel Pepys, 'talented, deserving boys were taught mathematics to enable them to become navigators in the English navy' (Jardine 2003: 61).

2 For earlier accounts which privilege the role of London over the universities in the development of modern science, see E.G.R. Taylor 1954, Christopher Hill 1965 and Webster 1975.

3 Mordechai Feingold describes Gresham as a failure, casting doubt upon the assumption of historians that an intended audience of practitioners were 'both capable of, and eager to understand' the output of their well-wishers (Feingold 1984: 176–7). He cites contemporary sources which complain both of the technical ignorance of scholars, and of the hostility to theoretically geared reform of practitioners, suggesting an absence of the necessary 'common ground' to justify an institution such as Gresham (Feingold 1984: 177). A. Rupert Hall argues that the literature of the Hartlib circle amounts to little more than 'gadgets' and get-rich-quick schemes, which neither increased technical proficiency through scientific information, nor made theoretical advances through technical experiment (A. Rupert Hall 1972: 49, 53). Hall's contention is that despite the meeting of ostensibly technicist and scientific minds, no meaningful interaction occurred.

4 Wren followed Sir John Denham, who followed John Webb, who followed Inigo Jones in the role of Surveyor-General of the Royal Works: effectively royal architect. William Petty oversaw the Down Survey of Irish estates forfeited after the rebellion of 1641. Jonas Moore was involved too in the great fire survey, but also in a range of other major mid-century projects including the 1650s drainage of the fens at Bedford Level; the 1661 Parliamentary survey of lands confiscated from the church; a 1662 survey of the Thames; and the building of a massive harbour wall, the Mole, at Tangier in 1663 (Jardine 2002: 171–3, 252–3).

5 See Matthew Stevenson's *Florus Britannicus* (1662) for a comparison between Archimedes and Henry VI, a pious yet ineffectual king (Stevenson 1662: 31).

6 Stephen Orgel drew attention to this emblem in his 1987 edition of *The Tempest*;

Peter Hulme and William Sherman have identified it as Archimedes (Shakespeare 1987: 21–2; Hulme and Sherman 2000: 9).

7 Thomas Randolph found in the association between Aristippus and drunkenness sufficient notoriety to justify an entire play (1630), whose long title reads: *Aristippvs, or The Iouiall Philosopher: Demonstrativelie Proouing, that Quartes, Pintes, and Pottles, are Sometimes Necessary Authours in a Scholers Library* (Randolph 1630).

8 Katherine Hill points out that this 'pose' of Oughtred's obscures his entanglement in the gift and salary economy of professional scholarship (Katherine Hill 1998: 259). It is no doubt reinforced by Aubrey, who has a tendency to tell tales of the Archimedean disinterest of mathematical scholars, even describing the materialist Hobbes as 'wont to draw lines on his thigh and on the sheetes, abed' (Aubrey 1972: 309).

9 See Edgerton 1975: 80–1 for an account of Alberti's painterly geometry.

10 *Mathematical Recreations* claims to be the translation of a French original by Jean Leurechon, and the British Library attributes the translation, together with some expansions, to William Oughtred. This attribution, which has a reasonably long history, has been disputed convincingly in a bibliographic study (Trevor Henry Hall 1969).

7 The doubtful traveller

1 John Gillies describes the physical comfort imagined in what he takes to be the always-implicit 'ultimate scene of cartography': the domestic interior 'in which maps are typically read' (Gillies 2001: 121).

2 See William Sherman's fascinating account of John Dee's 'living library' (Sherman 1995).

3 Steven Shapin certainly explores the symbolic aspect of the space of the experimental laboratory and movement within it, but is not particularly concerned with the rhetorical construction of this symbolic space through writing.

4 The printer's attribution of this poem to Denham is generally treated with scepticism, and it is often judged, along with the other poems printed with it, to be Marvell's.

5 See Schmidt 1997 for an account of Dutch supremacy in cartography in the seventeenth century.

6 Crystal Bartolovich has given an account of the 'advertising' function of early modern surveying tracts, which in her words use 'emblems, icons, and metaphors' in a struggle to effect a '"turn" from "feudally" to "mathematically" ordered spatial relations' (Bartolovich 1995: 255). Whilst I think this interpretation totalizes the politics of early modern surveying it is helpful in reminding us that usefulness was far from the only consideration in the promotion of an abstract, mathematical attitude to land.

7 J.A. Bennett describes manuals and the personae they construct as part of a wider 'propaganda effort' by mathematical practitioners to promote their arts, services and instruments (Bennett 1987: 38).

8 See Ravenhill 1983 for the long-running debate on whether Christopher Saxton triangulated or traversed.

9 A less colourful but perhaps more typical figure is Joshua Fisher, who performed ten surveys currently extant in the Massachusetts archives (Benes 1981: 36–7). Fisher was a member of the Dedham Church from 1639, a Representative for Dedham to the General Court, town clerk for four years, and a selectman for twenty-one. In Peter Benes' judgement, he was one of three men who 'virtually monopolized the town's affairs during the seventeenth century', and it was his 'position of trust and authority', rather than his mechanic skill or theoretical knowledge, that gave him the necessary weight to fix bounds between one town and another (Benes 1981: 37).

10 Both John Winthrop and William Penn became members of the Royal Society on the strength of their privileged access to America; Penn in 1681 directly as a result of producing Thomas Holme's map of Pennsylvania (Stearns 1946). Winthrop had sent a map to England as early as 1636 to answer the curiosity of antiquarian associates (Benes 1981: xvi).
11 See Benes 1981: 93, Figure 103, for a reproduction of and notes on Underhill's fort.
12 Hubbard wrote in response to King Philip's War (1676–77), a period of hostilities between the Wampanoag Indians and the New England colonies.
13 See Benes 1981: 93, Figure 9, for a reproduction of and notes on Foster's map.

Bibliography

Adams, I.H. (1976) *Agrarian Landscape Terms: a glossary for historical geography*, London: Institute of British Geographers.

Agas, Ralph (1596) *A Preparative to Plotting of Lands and Tenementes for Surveighs*, London: Thomas Scarlet.

—— (1606) *Note Touching Surveyors, and Against the Turning of Copyholds into Freeholds*, London: British Library Additional Ms. 12497.

Albano, Caterina (2001) 'Visible bodies: cartography and anatomy', in Andrew Gordon and Bernhard Klein (eds) *Literature, Mapping and the Politics of Space in Early Modern Britain*, Cambridge: Cambridge University Press, 89–106.

Alberti, Leon Battista (1972) *'On Painting' and 'On Sculpture'*, trans. Cecil Grayson, London: Phaidon.

Alexander, Amir (1995) 'The imperialist space of Elizabethan mathematics', *Studies in the History and Philosophy of Science*, 26: 559–92.

Allen, David Grayson (1982) *'Vacuum Domicilium*: the social and cultural landscape of seventeenth-century New England', in Jonathan L. Fairbanks and Robert F. Trent (eds) *New England Begins: the seventeenth century*, 3 vols, Boston: Museum of Fine Arts, I, 1–9.

Andrews, J.H. (1983) 'Appendix: the beginnings of the surveying profession in Ireland – abstract', in Sarah Tyacke (ed.) *English Map-Making 1500–1650: historical essays*, London: British Library, 20–1.

—— (2001) 'Introduction', in J.B. Harley and Paul Laxton, *The New Nature of Maps: essays in the history of cartography*, Baltimore and London: Johns Hopkins University Press, 2–32.

Anon. (1943) 'Essay on the ordering of towns', *Winthrop Papers*, 5 vols, Massachusetts Historical Society, III, Boston: Massachusetts Historical Society, 181–5.

Anton, Robert (1616) *The Philosophers Satyrs*, London: Roger Iackson.

Appelbaum, Robert (1998) 'Anti-geography', *Early Modern Literary Studies*, 4.2, Special Issue 3: 12.1–17.

Armitage, David (1998) 'Literature and empire', in Nicholas P. Canny (ed.) *The Oxford History of the British Empire, I: the origins of empire*, Oxford: Oxford University Press.

—— (2000) *The Ideological Origins of the British Empire*, Cambridge: Cambridge University Press.

Arneil, Barbara (1996) *John Locke and America: the defence of English colonialism*, Oxford: Clarendon.

Aubrey, John (1972) *Aubrey's 'Brief Lives'*, Harmondsworth: Penguin.

Aylett, Robert (1623) *Ioseph, or, Pharoah's Favourite*, London: M. Law.

Bacon, Francis, Viscount St Albans (1619) *The Wisedome of the Ancients*, trans. Sir Arthur Gorges, London: John Bill.

—— (2002) *The New Organon*, eds Lisa Jardine and Michael Silverthorne, Cambridge: Cambridge University Press.

Bancroft, Thomas (1639) *Two Bookes of Epigrammes and Epitaphs*, London: Matthew Walbancke.

Barber, Peter (1992) 'England II: monarchs, ministers, and maps, 1550–1625', in David Buisseret (ed.) *Monarchs, Ministers and Maps: the emergence of cartography as a tool of government in early modern Europe*, Chicago and London: University of Chicago Press.

Bartolovich, Crystal (1995) 'Spatial Stories: *The Surveyor* and the politics of transition', in Alvin Vos (ed.) *Place and Displacement in the Renaissance*, Binghampton: Center for Medieval and Early Renaissance Studies, State University of New York.

Basse, William (1893) *The Poetical Works of William Basse (1602–1653)*, ed. R. Warwick Bond, London: Ellis and Elvey.

Beaumont, Francis and Fletcher, John (1905–12) *Beaumont and Fletcher*, ed. Arnold Glover and A.R. Glover, Cambridge English Classics, 10 vols, London: Cambridge University Press.

Beaumont, Joseph (1880) *The Complete Poems of ... Joseph Beaumont, 1615–1699*, ed. Alexander B. Grosart, Blackburn: Chertsey Worthies Library.

Belyea, Barbara (1992) 'Images of power: Derrida/Foucault/Harley', *Cartographica*, 29.2: 1–9.

Bendall, Sarah A. (1996) 'Estate maps of an English county: Cambridgeshire, 1600–1830', in David Buisseret (ed.) *Rural Images: estate maps in the old and new worlds*, The Kenneth Nebenzahl, Jr. Lectures in the History of Cartography, Chicago and London: University of Chicago Press, 63–90.

Bending, Stephen and McRae, Andrew (eds) (2003) *The Writing of Rural England, 1500–1800*, Basingstoke and New York: Palgrave Macmillan.

Benes, Peter (ed.) (1981) *New England Prospect: a loan exhibition of maps at the Currier Gallery of Art*, Boston: Boston University Press.

Bennett, J.A. (1987) *The Divided Circle: a history of instruments for astronomy, navigation and surveying*, Oxford: Phaidon–Christie's Ltd.

Blith, Walter (1649) *The English Improver; or, a new survey of husbandry*, London: John Wright.

—— (1652) *The English Improver Improved; or, the survey of husbandry surveyed*, London: John Wright.

Boelhower, William (1988) 'Inventing America: a model of cartographic semiosis', *Word and Image*, 4.2: 475–97.

Bowen, Clarence Winthrop (1882) *The Boundary Disputes of Connecticut*, Boston: James R. Osgood.

Bradford, William (1946) *Bradford's History of Plymouth Plantation 1606–1646*, ed. William T. Davis, Original Narratives of Early American History, New York: Barnes and Noble.

Brenner, Robert (1985) 'Agrarian class structure and economic development in pre-industrial Europe', in T.H. Aston and C.H.E. Philpin (eds) *The Brenner Debate: Agrarian Class Structure and Economic Development in Pre-Industrial Europe*, Cambridge: Cambridge University Press.

Brome, Richard (1968) *A Joviall Crew*, ed. Anne Haaker, London: Edward Arnold.

Brotton, Jerry (1997) *Trading Territories: mapping the early modern world*, London: Reaktion.

Browne, William (1868–69) *The Whole Works of William Browne, of Tavistock*, ed. W. Carew Hazlitt, 2 vols, London: Roxburghe Library.

Buisseret, David (1996a) 'Introduction: defining the estate map', in David Buisseret (ed.) *Rural Images: estate maps in the old and new worlds*, The Kenneth Nebenzahl, Jr. Lectures in the History of Cartography, Chicago and London: University of Chicago Press, 1–4.

—— (1996b) 'The estate map in the new world', in David Buisseret (ed.) *Rural Images: estate maps in the old and new worlds*, The Kenneth Nebenzahl, Jr. Lectures in the History of Cartography, Chicago and London: University of Chicago Press, 91–112.

Bullough, Geoffrey (1939) 'Introduction', in Fulke Greville, Baron Brooke, *Poems and Dramas of Fulke Greville, First Lord Brooke*, ed. Geoffrey Bullough, 2 vols, Edinburgh: Oliver and Boyd, I, 1–72.

Burt, Richard and Archer, John Michael (1994) 'Introduction', in Richard Burt and John Michael Archer (eds) *Enclosure Acts: sexuality, property, and culture in early modern England*, Ithaca and London: Cornell University Press, 1–13.

Burton, Robert (1893) *The Anatomy of Melancholy*, ed. Rev. A.R. Shilleto, 3 vols, London: Bell.

Bushman, R.L. (1998) 'Markets and composite farms in early America', *William and Mary Quarterly*, 55.3: 351–74.

Butler, Martin (2004) 'Brome, Richard (*c*.1590–1652)', *Oxford Dictionary of National Biography*, Oxford University Press. Available online at: http://www.oxforddnb.com/view/article/3503 (accessed 22 October 2004).

Butlin, R.A. (1979) 'The enclosure of open fields and extinction of common rights in England, *circa* 1600–1750: a review', in H.S.A. Fox and R.A. Butlin (eds) *Change in the Countryside: essays on rural England, 1500–1900*, Institute of British Geographers Special Publication 10, London: Institute of British Geographers, 65–82.

Candee, Richard M. (1982) 'Land surveys of William and John Godsoe of Kittery, Maine: 1689–1769', in Peter Benes (ed.) *New England Prospect: maps, place names, and the historical landscape*, Boston: Boston University Press, 9–46.

Canny, Nicholas P. (1988) '"To Establish a Common Wealthe": Captain John Smith as new world colonist', *Virginia Magazine of History and Biography*, 92.2: 213–22.

Carew, Thomas (1949) *The Poems of Thomas Carew with his Masque, 'Coelum Britannicum'*, ed. Rhodes Dunlop, Oxford: Clarendon.

Carroll, Peter N. (1969) *Puritanism and the Wilderness: the intellectual significance of the New England frontier 1629–1700*, New York and London: Columbia University Press.

Cassirer, Ernst (1970) *The Platonic Renaissance in England*, trans. James P. Pettegrove, New York: Gordian Press.

Cavendish, Margaret, Duchess of Newcastle (1668) *Plays, Never before Printed*, 5 vols, London: A. Maxwell.

Chaplin, Joyce (2001) *Subject Matter: technology, the body, and science on the Anglo-American frontier, 1500–1676*, Cambridge, Massachusetts: Harvard University Press.

Clucas, Stephen (2000) 'Thomas Harriot and the field of knowledge in the English Renaissance', in Robert Fox (ed.) *Thomas Harriot: an Elizabethan man of science*, Aldershot: Ashgate, 93–136.

Collinson, Patrick (1996) 'Elizabethan and Jacobean Puritanism as forms of popular religious culture', in Christopher Durston and Jacqueline Eales (eds) *The Culture of English Puritanism 1560–1700*, Basingstoke: Macmillan, 32–57.

Conley, Tom (1996) *The Self-Made Map: cartographic writing in early modern France*, Minneapolis and London: University of Minnesota Press.

Corbet, Richard (1871) *The Times' Whistle; or, a newe daunce of seuen satires, and other poems*, ed. J.M. Cowper, Early English Text Society Original Series 46, London: N. Trubner and Co.

Corcoran, Irma (1992) *Thomas Holme, 1624–1695: Surveyor General of Pennsylvania*, Philadelphia: American Philosophical Society.

Cormack, Lesley B. (1991) '"Good Fences Make Good Neighbors": geography as self-definition in early modern England', *Isis*, 82: 639–61.

—— (1997) *Charting an Empire: Geography at the English Universities, 1580–1620*, Chicago: University of Chicago Press.

—— (2001) 'Britannia rules the waves? Images of empire in Elizabethan England', in Andrew Gordon and Bernhard Klein (eds) *Literature, Mapping and the Politics of Space in Early Modern Britain*, Cambridge: Cambridge University Press.

Cosgrove, Denis (1993) *The Palladian Landscape: geographical change and its cultural representation in sixteenth-century Italy*, Leicester: Leicester University Press.

—— (1999) 'Introduction: mapping meaning', in Denis Cosgrove (ed.) *Mappings*, London: Reaktion, 1–23.

—— and Daniels, Stephen (1988) 'Introduction: iconography and landscape', in Denis Cosgrove and Stephen Daniels (eds) *The Iconography of Landscape: essays on the symbolic representation, design and use of past environments*, Cambridge Studies in Historical Geography 9, Cambridge: Cambridge University Press, 1–10.

Cowley, Abraham (1905–06) *The Complete Works in Verse and Prose of Abraham Cowley*, ed. Alexander B. Grosart, 2 vols, New York: AMS Press inc. and Cambridge: Cambridge University Press.

Crang, Mike (1998) *Cultural Geography*, London and New York: Routledge.

—— (2000) 'Relics, places and unwritten geographies in the work of Michel de Certeau (1925–86)', in Mike Crang and Nigel Thrift (eds) *Thinking Space*, London and New York: Routledge, 136–53.

Crossley, E.W. and Willan, T.S. (eds) (1941) *Three Seventeenth-Century Yorkshire Surveys*, Leeds: Yorkshire Archæological Society, 104.

Cumming, William P. (1972) *British Maps of Colonial America*, The Kenneth Nebenzahl, Jr. Lectures in the History of Cartography, London and Chicago: Chicago University Press.

Cuningham, William (1559) *The Cosmographical Glasse*, London: John Daye.

Darby, H.C. (1973) 'The Age of the Improver: 1600–1800', in H.C. Darby (ed.) *A New Historical Geography of England After 1600*, Cambridge: Cambridge University Press, 1–88.

Davenport, Robert (1661) *The City-Night-Cap*, London: Samuel Speed.

de Certeau, Michel (1984) *The Practice of Everyday Life*, trans. Steven F. Rendall, Berkeley: University of California Press.

Dee, John (1570) 'The Mathematicall Praeface', annexed to Euclid, *The Elements of Geometrie of … Euclide*, trans. Sir Henry Billingsley, London: John Daye.

Defoe, Daniel (1976) *The Fortunes and Misfortunes of the Famous Moll Flanders, &c.*, ed. G.A. Starr, London and Oxford: Oxford University Press.

Denham, Sir John and Marvell, Andrew (1667) *Directions to a Painter for Describing our Naval Business in Imitation of Mr Waller … Whereunto is Annexed, Clarindons Housewarming, by an Unknown Author*, London.

Derrida, Jacques (1978) *Edmund Husserl's 'Origin of Geometry': an introduction*, trans. John P. Leavey, ed. David B. Allison, New York: N. Hays and Hassocks: Harvester.

Digges, Leonard and Digges, Thomas (1571) *Pantometria*, London: Henrie Bynneman.

—— (1579) *An Arithmeticall Militare Treatise, Named Stratioticos …*, London: Henrie Bynneman.

Du Bartas, Guillaume de Salluste, seigneur (1979) *The Divine Weeks and Works of Guillaume de Saluste, sieur Du Bartas, translated by Josuah Sylvester*, trans. Josuah Sylvester, ed. Susan Snyder, 2 vols, Oxford: Clarendon.

Du Jon, François (1638) *The Painting of the Ancients*, London: Richard Hodgkinsonne.

Durston, Christopher and Eales, Jacqueline (1996) 'Introduction: the Puritan ethos, 1560–1700', in Christopher Durston and Catherine Eales (eds) *The Culture of English Puritanism 1560–1700*, Basingstoke: Macmillan, 1–31.

Dymock, Cressey (1653) *A Discovery for New Divisions; or, setting out of lands, as to the best forme: imparted in a letter to Samuel Hartlib*, in Samuel Hartlib (ed.) *A Discoverie For Division Or Setting out of Land, as to the Best Form … whereunto are added some other choice secrets or experiments of husbandry*, London: Richard Wodnothe.

Edgerton, Samuel (1975) *The Renaissance Rediscovery of Linear Perspective*, New York: Basic Books.

—— (1983) 'From mental matrix to *mappamundi* to Christian empire: the heritage of Ptolemaic cartography in the renaissance', in David Woodward (ed.) *Art and Cartography: six historical essays*, Chicago: University of Chicago Press, 10–50.

Edney, Matthew (1993) 'Cartography without "progress"', *Cartographica*, 30.2/3: 54–68.

—— (1996) 'Theory and the history of cartography', *Imago Mundi*, 48: 185–91.

—— and Cimburek, Susan (2004) 'Telling the traumatic truth: William Hubbard's narrative of King Philip's war and his "Map of New-England"', *William and Mary Quarterly*, 61.2: 317–48.

Edwards, A.C. and Newton, K.C. (1984) *The Walkers of Hanningfield: surveyors and map-makers extraordinary*, London: Buckland.

Euclid (1703) *Ευκλειδου τα σωζομενα. Euclidis quæ supersunt omnia* ..., trans. David Gregory, Oxford: Theatro Sheldoniano.

Fairbanks, Jonathan L. and Trent, Robert F. (eds) (1982) *New England Begins: the seventeenth century*, 3 vols, Boston: Museum of Fine Arts.

Farley, Robert (1638) *Kalendarium Humanæ Vitæ: the kalender of man's life*, London: W. Hope.

Feingold, Mordechai (1984) *The Mathematician's Apprenticeship: science, universities and society in England, 1560–1640*, Cambridge: Cambridge University Press.

Fitzherbert (1523) *Here Begynneth a Ryght Frutefull Mater, and Hath to Name the Boke of Surueyeng and Improumentes*, London: Rycharde Pynson.

Folkingham, William (1610) *Feudigraphia: the synopsis or epitome of surveying methodized*, London: R. Moore.

Fordham, Sir H.G. (1929) *Some Notable Surveyors and Map-Makers of the Sixteenth, Seventeenth, and Eighteenth Centuries and their Work: a study in the history of cartography*, Cambridge: Cambridge University Press.

Foucault, Michel (1980) 'Questions on geography: interview with the editors of *Hérodote*', in Colin Gordon (ed.) *Power/Knowledge: selected interviews and other writings 1972–1977 by Michel Foucault*, Brighton: Harvester Press, 63–77.

—— (1986a) 'Panopticism', in Paul Rabinow (ed.) *The Foucault Reader*, Harmondsworth: Penguin, 206–13.

—— (1986b) 'Space, Knowledge, and Power', in Paul Rabinow (ed.) *The Foucault Reader*, Harmondsworth: Penguin, 239–56.

Fowler, William (1914–40) *The Works of William Fowler, secretary to Queen Anne, wife of James VI*, ed. Henry W. Meikle, 3 vols, Edinburgh: W. Blackwood and Sons.

Franklin, Wayne (1979) *Discoverers, Explorers, Settlers: the diligent writers of early America*, Chicago, IL and London: University of Chicago Press.

French, Peter (1972) *John Dee: the world of an Elizabethan magus*, London: Routledge.

Fuller, Mary C. (1995) *Voyages in Print*, Cambridge: Cambridge University Press.

Garden, Alexander (1625) *Characters and Essayes*, Aberdeen: Edward Raban.

Gerbier, Balthazar (1649) *The First Lecture of Geographie, (which is a Description of the Terrestriall Globe) Read Publickly at Sr B. Gerbier his Academy at Bednall-Greene*, London: Hanna Allen.

Gerish, William Blyth (1903) *John Norden, 1548–1626 (?): a biography*, Ware: G. Price & Son.

Gillies, John (1994) *Shakespeare and the Geography of Difference*, Cambridge: Cambridge University Press.

—— (2001) 'The scene of cartography in *King Lear*', in Andrew Gordon and Bernhard Klein (eds) *Literature, Mapping and the Politics of Space in Early Modern Britain*, Cambridge: Cambridge University Press, 109–37.

Gonner, E.C.K. (1912) *Common Land and Enclosure*, London: Macmillan.

Gosson, Philip (1579) *The Ephemerides of Phialo*, London: T. Dawson.

Gouws, John (2004) 'Greville, Fulke, first Baron Brooke of Beauchamps Court (1554–1628)', *Oxford Dictionary of National Biography*, Oxford University Press. Available online at: http://www.oxforddnb.com/view/article/11516 (accessed 22 October 2004).

Grasso, Christopher (1999) 'Review: Stephen Innes, *Creating the Commonwealth: The economic culture of Puritan New England*', *William and Mary Quarterly*, 56.1, 201–2.

Greenblatt, Stephen (1981) 'Invisible Bullets: Renaissance authority and its subversion, *Henry IV* and *Henry V*', *Glyph*, 8: 40–61; also collected in Jonathan Dollimore and

Alan Sinfield (eds) (1985) *Political Shakespeare: new essays in cultural materialism*, Manchester: Manchester University Press, 18–47.

Greville, Fulke, Baron Brooke (1939) *Poems and Dramas of Fulke Greville, First Lord Brooke*, ed. Geoffrey Bullough, 2 vols, Edinburgh: Oliver and Boyd.

—— (1965) *The Remains*, ed. G.A. Wilkes, London: Oxford University Press.

Gronim, Sara Stidstone (2001) 'Geography and persuasion: maps in British colonial New York', *William and Mary Quarterly*, 58.2: 373–402.

Grotius, Hugo (1652) *Hugo Grotius, His Sophompaneas, or Ioseph, a Tragedy*, trans. Francis Goldsmith, London: John Hardesty.

Haak, Theodore, Hartlib, Samuel, Oldenburg, Henry and Winthrop, John (1878) *Correspondence of Hartlib, Haak, Oldenburg, and Others of the Founders of the Royal Society, with Governor Winthrop of Connecticut, 1661–1672*, Boston: John Wilson and Son.

Hagthorpe, John (1623) *Visiones Rerum: The visions of things, or foure poems*, London: Bernard Alsop.

Hainsworth, D.R. (1992) *Stewards, Lords, and People: the estate steward and his world in later Stuart England*, Cambridge Studies in Early Modern British History, Cambridge: Cambridge University Press.

Hakluyt, Richard (1903–05) *The Principal Navigations, Voyages, Traffiques & Discoveries of the English Nation*, 12 vols, Glasgow: James MacLehose and Sons.

—— and Hakluyt, Richard (1935) *The Original Writings & Correspondence of the two Richard Hakluyts*, ed. E.G.R. Taylor, Hakluyt Society Second Series 76 and 77, 2 vols, London: Hakluyt Society.

Hall, A. Rupert (1972) 'Science, technology and utopia in the seventeenth century', in Peter Mathias (ed.) *Science and Society 1600–1900*, Cambridge: Cambridge University Press, 33–53.

Hall, David D. (1989) *Worlds of Wonder, Days of Judgment: popular religious belief in early New England*, New York: Random House.

Hall, Trevor Henry (1969) *'Mathematical Recreations': an exercise in seventeenth-century bibliography*, Leeds Studies in Bibliography and Textual Criticism Occasional Paper 1, Leeds: Leeds University School of English.

Harley, J.B. (1988) 'Maps, knowledge, and power', in Denis Cosgrove and Stephen Daniels (eds) *The Iconography of Landscape: essays on the symbolic representation, design and use of past environments*, Cambridge Studies in Historical Geography 9, Cambridge: Cambridge University Press, 277–312; also collected in J.B. Harley (2001) *The New Nature of Maps: essays in the history of cartography*, ed. Paul Laxton, Baltimore: Johns Hopkins University, 52–81.

—— (2001) 'New England cartography and the Native Americans', in J.B. Harley, *The New Nature of Maps: essays in the history of cartography*, ed. Paul Laxton, Baltimore: Johns Hopkins University Press, 169–95; first published in Emerson Baker, Edwin A. Churchill, Richard S. D'Abate, Kristine L. Jones, Victor A. Konrad and Harald E.L. Prins (eds) (1994) *American Beginnings: exploration, culture, and cartography in the land of Norumbega*, Lincoln: University of Nebraska Press, 287–313.

Harrington, James (1992) *'The Commonwealth of Oceana' and 'System of Politics'*, ed. J.G.A. Pocock, Cambridge Texts in the History of Political Thought, Cambridge: Cambridge University Press.

Harriot, Thomas (1972) *A Briefe and True Report of the New Found Land of Virginia: the complete 1590 Theodor de Bry edition*, New York: Dover.

Harris, John, Orgel, Stephen and Strong, Roy (1973) *The King's Arcadia: Inigo Jones and the Stuart court*, London: Arts Council of Great Britain.

Hartlib, Samuel (1653) *A Discoverie For Division Or Setting out of Land, as to the Best Form ... Whereunto are Added some other Choice Secrets or Experiments of Husbandry ...*, London: Richard Wodnothe.

Harvey, David (1989) *The Condition of Postmodernity: an enquiry into the origins of cultural change*, Oxford: Basil Blackwell.

Harvey, P.D.A. (1980) *The History of Topographical Maps: symbols, pictures and surveys*, London: Thames and Hudson.

—— (1996) 'English estate maps: their early history and their use as historical evidence', in David Buisseret (ed.) *Rural Images: Estate Maps in the Old and New Worlds*, The Kenneth Nebenzahl, Jr., Lectures in the History of Cartography, Chicago and London: University of Chicago Press, 27–61.

Hayes, Kevin J. (1991) 'Defining the ideal colonist: captain John Smith's revisions from *A True Relation to the Proceedings* to the third book of the *General Historie*', *Virginia Magazine of History and Biography*, 99.2: 123–44.

Heath, Robert (1650) *Epigrams*, London: H. Moseley.

Helgerson, Richard (1992) *Forms of Nationhood: the Elizabethan writing of England*, Chicago and London: University of Chicago Press.

—— (1993) 'Nation or estate: ideological conflict in the early modern mapping of England', *Cartographica*, 30: 68–74.

—— (1998) 'Introduction', *Early Modern Literary Studies*, 4.2, Special Issue 3: 1.1–14.

—— (2001) 'The folly of maps and modernity', in Andrew Gordon and Bernhard Klein (eds) *Literature, Mapping and the Politics of Space in Early Modern Britain*, Cambridge: Cambridge University Press, 241–62.

Herodotus (1924) *The Famous Hystory of Herodotus*, trans. Barnabe Rich, The Tudor Translations Second Series 6, London: Constable.

Hill, Christopher (1965) *Intellectual Origins of the English Revolution*, Oxford: Clarendon.

—— (1980, 2nd edn; first published 1961) *The Century of Revolution 1603–1714*, Walton-on-Thames: Nelson.

Hill, Katherine (1998) '"Juglers or schollers": negotiating the role of a mathematical practitioner', *The British Journal for the History of Science*, 31: 253–74.

Hood, Thomas (1974) *A Copie of the Speache Made by The Mathematicall Lecturer ... at the House of Thomas Smith*, Amsterdam: Theatrum Orbis Terrarum.

Hooykaas, R. (1958) *Science, Humanisme, et Réforme: Pierre de la Ramée, 1515–1572*, Leyde: E.J. Brill.

Howson, Geoffrey (1982) *A History of Mathematical Education in England*, Cambridge: Cambridge University Press.

Hubbard, William (1677) *The Present State of New England, being a Narrative of the Troubles with the Indians*, London: Thomas Parkhurst.

Huggan, Graham (1989) 'Decolonizing the map: post-colonialism, post-structuralism and the cartographic connection', *Ariel: A Review of International English Literature*, 20.4: 115–31.

Hulme, Peter (1981) 'Hurricanes in the Caribbees: the constitution of the discourse of English colonialism', in Francis Barker *et al.* (eds) *1642: literature and power in the seventeenth century*, Colchester: University of Essex, 55–83.

—— (1986) *Colonial Encounters: Europe and the Native Caribbean 1492–1797*, London: Methuen.

—— and Sherman, William H. (2000) 'Local knowledge: introduction', in Peter Hulme and William H. Sherman (eds) *'The Tempest' and its Travels*, London: Reaktion, 3–11.

Innes, Stephen (1995) *Creating the Commonwealth: the economic culture of Puritan New England*, New York and London: Norton.

Ive, Paul (1589) *The Practise of Fortification*, London: T. Man and T. Cooke.

Jackson, Peter (1989) *Maps of Meaning: an introduction to cultural geography*, London: Unwin Hyman.

James I (1618) *The Kings Maiesties Declaration to his Subiects, Concerning Lawfull Sports to be Vsed*, London: Bonham Norton and Iohn Bill.

Jardine, Lisa (2002) *On a Grander Scale: the outstanding career of Christopher Wren*, London: HarperCollins.

—— (2003) *The Curious Life of Robert Hooke*, London: HarperCollins.

—— and Sherman, William (1994) 'Pragmatic readers: knowledge transactions and scholarly services in late Elizabethan England', in Anthony Fletcher and Peter Roberts (eds) *Religion, Culture and Society in Early Modern Britain: essays in honour of Patrick Collinson*, Cambridge: Cambridge University Press, 102–24.

Jehlen, Myra (1993) 'Why did the Europeans cross the ocean?', in Amy Kaplan and Donald Pease (eds) *Cultures of United States Imperialism*, Duke University Press, 41–58.

Jennings, Francis (1975) *The Invasion of America: Indians, colonialism, and the cant of conquest*, Chapel Hill: University of North Carolina Press.

Jesseph, Douglas M. (1999) *Squaring the Circle: the war between Hobbes and Wallis*, Chicago, IL: University of Chicago Press.

Johnson, Edward (1974) *Wonder-working Providence of Sion's Saviour in New-England, 1654, and Good News from New England, 1648*, Delmar, New York: Scholars' Facsimiles & Reprints.

Johnson, Hildegard Binder (1976) *Order Upon the Land: the U.S. rectangular land survey and the upper Mississippi country*, London and New York: Oxford University Press.

Jonson, Ben (1925–52) *Ben Jonson*, ed. C.H. Herford, Percy Simpson and Evelyn Simpson, 11 vols, Oxford: Clarendon.

—— (1969) *The Complete Masques*, ed. Stephen Orgel, New Haven and London: Yale University Press.

Josselyn, John (1988) *John Josselyn, Colonial Traveller: a critical edition of 'Two Voyages to New-England'*, ed. Paul J. Lindholdt, Hanover, New Hampshire and London: University Press of New England.

Kearney, H.F. (1973) 'Merton revisited', *Science Studies*, 3.1: 72–8.

Kerridge, Eric (1969) *Agrarian Problems in the Sixteenth Century and After*, Historical Problems: Studies and Documents 6, London: Allen & Unwin.

Kettle, Arnold (1988) *Literature and Liberation: selected essays*, ed. G. Martin and W.R. Owens, Manchester: Manchester University Press.

Kitchen, Frank (2004) 'Norden, John (*c.*1547–1625)', *Oxford Dictionary of National Biography*, Oxford University Press. Available oneline at: http://www.oxforddnb.com/view/article/20250 (accessed 22 October 2004).

Klein, Bernhard (2001) *Maps and the Writing of Space in Early Modern England and Ireland*, London: Palgrave.

Knapp, Jeffrey (1992) *An Empire Nowhere: England, America, and literature from 'Utopia' to 'The Tempest'*, Berkeley and London: University of California Press.

Kolodny, Annette (1975) *The Lay of the Land: metaphor as experience and history in American life and letters*, Chapel Hill: University of North Carolina Press.

Koyre, Alexander (1957) *From the Closed World to the Infinite Universe*, Baltimore: Johns Hopkins University Press.

Krim, Arthur J. (1982) 'Acculturation of the New England landscape: native and English toponymy of eastern Massachusetts', in Peter Benes (ed.) *New England Prospect: maps, place names, and the historical landscape*, The Dublin Seminar for New England Folklife: Annual Proceedings, Boston: Boston University Press, 69–88.

Kulikoff, Allan (2000) *From British Peasants to Colonial American Farmers*, Chapel Hill: University of North Carolina Press.

Kupperman, Karen Ordahl (1980) *Settling with the Indians: the meeting of English and Indian cultures in America, 1580–1640*, London: Dent.

—— (1997) 'Presentment of civility: English reading of American self-presentation in the early years of colonization', *William and Mary Quarterly*, 54.1: 193–228.

Laqueur, Thomas (1990) *Making Sex: body and gender from the Greeks to Freud*, Cambridge, Massachusetts: Harvard University Press.

Laird, W.R. (1991) 'Archimedes among the humanists', *Isis*, 82: 629–38.

Lee, Joseph (1656) *Ευταξια του Αγρου, or a vindication of a regulated inclosure*, London.

Lefebvre, Henri (1991) *The Production of Space*, trans. Donald Nicholson-Smith, Oxford: Blackwell.

Leigh, Valentine (1562) *The Most Profitable and Commendable Science of Surueying of Lands, Tenements, and Hereditaments ...*, London: Robert Dexter.

—— (1577) *The Moste Profitable and Commendable Science, of Surueying of Landes, Tenementes, and Hereditamentes ...*, London: Miles Jennings.

Leslie, Michael and Raylor, Timothy (1992) 'Introduction', in Michael Leslie and Timothy Raylor (eds) *Culture and Cultivation in Early Modern England: writing and the land*, Leicester: Leicester University Press.

Lewis, G. Malcolm (1998) 'Introduction', in G. Malcolm Lewis (ed.) *Cartographic Encounters: perspectives on native American mapmaking and map use*, Chicago: University of Chicago Press, 1–6.

Leybourn, William (1653) *The Compleat Surveyour*, London: E. Brewster and G. Sawbridge.

—— (1667) *The Line of Proportion or Numbers Commonly Called Gunters Line, Made Easie*, London: G. Sawbridge.

—— (1673, 2nd edn; first published 1667) *The Line of Proportion or Numbers Commonly called Gunters Line, Made Easie*, London: G. Sawbridge.

—— (1690) *Cursus Mathematicus*, London: Thomas Basset, Benjamin Tooke, Thomas Sawbridge, Awnsham and John Churchill.

Locke, John (1967, 2nd edn; first published 1960) *Two Treatises of Government*, ed. Peter Laslett, Cambridge: Cambridge University Press.

—— (1975) *An Essay Concerning Human Understanding*, ed. Peter H. Nidditch, Oxford: Clarendon.

Love, John (1688) *Geodaesia; or, the Art of Surveying and Measuring of Land Made Easie*, London: John Taylor.

—— (1715, 2nd edn; first published 1688) *Geodaesia; or, the Art of Surveying and Measuring of Land Made Easie*, London: W. Taylor.

Lucar, Cyprian (1590) *Lucarsolace*, London: Iohn Harrison.

Lupton, Julia (1993) 'Mapping mutability; or, Spenser's Irish plot', in Brendan Bradshaw, Andrew Hadfield and Willy Maley (eds) *Representing Ireland: literature and the origins of conflict, 1534–1660*, Cambridge: Cambridge University Press, 93–115.

McHugh, P.G. (1999) 'A Tribal Encounter: the presence and properties of common law language in the discourse of colonisation in the early-modern period', in Alex Calder, Jonathan Lamb and Bridget Orr (eds) *Voyages and Beaches: Pacific Encounters, 1769–1840*, Honolulu: University of Hawai'i Press, 114–31.

Mackenthun, Gesa (1997) *Metaphors of Dispossession: American beginnings and the translation of empire, 1492–1637*, Norman: University of Oklahoma Press.

McKeon, Michael (1987) *Origins of the English Novel, 1600–1740*, Baltimore: Johns Hopkins University Press.

McLeod, Bruce (1999) *The Geography of Empire in English Literature, 1580–1745*, Cambridge and New York: Cambridge University Press.

McRae, Andrew (1992) 'Husbandry manuals and the language of agricultural improvement', in Michael Leslie and Timothy Raylor (eds) *Culture and Cultivation in Early Modern England: writing and the land*, Leicester: Leicester University Press, 35–62.

—— (1996) *God Speed the Plough: the representation of agrarian England, 1500–1660*, Cambridge: Cambridge University Press.

Manlius, Marcus and Sherburne, Sir Edward (1675) *The Sphere of Marcus Manilius made an English Poem*, London: Nathanael Brooke.

Markham, Gervase (1609) *The Famous Whore, or Noble Curtizan*, London: J. Budge.

Marmion, Shackerley (1875) *The Dramatic Works of Shackerley Marmion*, Edinburgh: William Paterson and London: H. Sotheran and Co.

Marolois, Samuel (1638) *The Art of Fortification*, trans. Henry Hexham, Amsterdam.

Martial and Fletcher, R. (1656) *Ex Otio Negotium; or, Martiall his epigrams translated*, trans. R. Fletcher, London: William Shears.

Marvell, Andrew (1972) *The Complete Poems*, ed. Elizabeth Story Donno, 2 vols, Harmondsworth: Penguin.

Mason, John (1897) 'A brief history of the Pequot war', in Charles Orr (ed.) *History of the Pequot War: the contemporary accounts of Mason, Underhill, Vincent and Gardener*, Cleveland: Helman-Taylor Co.

Mather, Increase (1676) *A Brief History of the Warr with the Indians in New England*, London: R. Chirwell.

Merton, Robert K. (1938) 'Science, technology and society in seventeenth-century England', *Osiris*, 4: 360–632.

Middleton, Thomas (1885–86) *Works*, ed. A.H. Bullen, 8 vols, London: John C. Nimmo.

Miller, Perry (1956) *Errand into the Wilderness*, Cambridge, Massachusetts: Harvard University Press and London: Oxford University Press.

Milton, John (1953–82) *Complete Prose Works of John Milton*, ed. Don M. Wolfe, Ernest Sirluck, Merritt Y. Hughes, Maurice Kelley and Robert W. Ayers, 8 vols, New Haven: Yale University Press and London: Oxford University Press.

Montrose, Louis (1991) 'The work of gender in the discourse of discovery', *Representations*, 33 Winter: 1–41; also collected in Stephen Greenblatt (ed.) *New World Encounters*, Berkeley: University of California Press, 197–217.

Moore, Adam (1653) *Bread for the Poor and Advancement of the English Nation Promised by Enclosure of the Wastes and Common Grounds of England*, London: Nicholas Bourn.

Moore, John (1653) *The Crying Sin of England, of Not Caring for the Poor …*, London: Antony Williamson.

Moore, Sir Jonas (1685) *The History or Narrative of the Great Level of the Fenns Called Bedford Level*, London: Moses Pitt.

More, Henry (1647) *Philosophicall Poems*, Cambridge: Roger Daniel.

Morrison, Russell, Papenfuse, E., Bramucci, N. and Janson-La Palme, R. (eds) (1983) *On The Map: an exhibit and catalogue of maps relating to Maryland and the Chesapeake Bay …*, Chestertown: Washington College.

Nash, Gary B. (1986) *Race, Class, and Politics: essays on American colonial and revolutionary society*, Urbana: University of Illinois Press.

Norden, John (1593) *Speculum Britanniae: the First Parte. An Historical and Chorographicall Discription of Middlesex*, London: Eliot's Court Press.

—— (1596) *Preparative to His Speculum Britanniae*, London: John Windet.

—— (1607) *The Surveyor's Dialogue*, London: Hugh Astley.

—— (1617) *Survey of Prince Charles's Manors, et cetera*, London: British Library Additional Ms. 6027.

—— (1931) *Vicissitudo Rerum*, Shakespeare Association Facsimiles 4, London: Oxford University Press for the Shakespeare Association.

Nuti, Lucia (1999) 'Mapping places: chorography and vision in the Renaissance', in Denis Cosgrove (ed.) *Mappings*, London: Reaktion, 90–108.

Ogborn, Miles (1998) *Spaces of Modernity: London's geographies, 1680–1780*, London and New York: Guilford Press.

Orgel, Stephen (1965) *The Jonsonian Masque*, London: Oxford University Press and Cambridge, Massachusetts: Harvard University Press.

Oughtred, William (1631) *Clavis Mathematicae*, London: Thomas Harper.

—— (1632) *The Circles of Proportion and the Horizontal Instrument. Both inuented, and the vses of both written in Latine by Mr. W.O.*, trans. William Forster, London: Elias Allen.

—— (1633) *Mathematical Recreations*, London: Richard Hawkins.

Ovid (1961) *Shakespeare's Ovid, Being Arthur Golding's Translation of the Metamorphoses edited by W.H.D. Rouse, Litt. D.*, trans. Arthur Golding, ed. W.H.D. Rouse, London: Centaur; reduced photographic reprint of 1904 edition, published London: De La More Press.

—— (1970) *Ovid's Metamorphosis Englished, Mythologized, and Represented in Figures*, trans. George Sandys, Lincoln: University of Nebraska Press.

Pagden, Anthony (1995) *Lords of all the World: ideologies of empire in Spain, Britain and France c.1500–c.1800*, New Haven and London: Yale University Press.

Panofsky, Erwin (1955) *Meaning in the Visual Arts: papers in and on art history*, New York: Doubleday.

Peacham, Henry (1612) *Minerva Britanna; or, a Garden of Heroical Deuises, Furnished, and Adorned with Emblemes and Impresa's of Sundry Natures*, London: Walter Dight.

Pearce, Margaret Wickens (1988) 'Native mapping in southern New England Indian deeds', in G. Malcolm Lewis (ed.) *Cartographic Encounters: perspectives on Native American mapmaking and map use*, Chicago: University of Chicago Press, 157–86.

Pell, John (1650) 'An idea of mathematics written by Mr John Pell to Samuel Hartlib', in John Dury and Samuel Hartlib (eds) *The Reformed Librarie-Keeper*, London.

Penrose, Boies (1952) *Travel and Discovery in the Renaissance, 1420–1620*, Cambridge, Massachusetts: Harvard University Press.

Peterson, M.A. (1997) *The Price of Redemption: the spiritual economy of Puritan New England*, Stanford: Stanford University Press.

—— (2001) 'Puritanism and refinement in early New England: reflections on communion silver', *William and Mary Quarterly*, 58.2: 307–46.

Pettie, George (1576) *A Petite Pallace of Pettie his Pleasure: contayning many pretie hystories by him set foorth*, London: R. Watkins.

Plattes, Gabriel (1644) *The Profitable Intelligencer, Communicating his Knowledge for the Generall Good of the Common-wealth and all Posterity …*, London.

—— (1979) 'A description of the famous kingdome of Macaria', in Charles Webster, *Utopian Planning and the Puritan Revolution: Gabriel Plattes, Samuel Hartlib, and Macaria*, Oxford: Research Publications of the Wellcome Unit for the History of Medicine.

Plutarch (1595, 2nd edn; first published 1579) *The Lives of the Ancient Grecians and Romans, Compared Together*, trans. Sir Thomas North, London: Thomas Wight.

Pocock, C.D. (1988) 'Interface: geography and literature', *Progress in Human Geography*, 12: 87–102.

Pocock, J.G.A. (1975) *The Machiavellian Moment: Florentine political thought and the Atlantic republican tradition*, Princeton, New Jersey: Princeton University Press.

Powell, Robert (1636) *Depopulation Arraigned, Convicted and Condemned, by the Lawes of God and Man*, London: Richard Badger.

Pratt, Mary Louise (1992) *Imperial Eyes: travel writing and transculturation*, London: Routledge.

Prothero, Rowland Edmund, Baron Ernle (1961, 6th edn; first published 1912) *English Farming Past and Present*, London: Heineman.

Purchas, Samuel (ed.) (1905–07) *Hakluytus Posthumus; or, Purchas His Pilgrimes*, 20 vols, Glasgow: James MacLehose & Sons.

Quarles, Francis (1967) *The Complete Works*, ed. Alexander B. Grosart, Chertsey Worthies Library, 3 vols, New York: A.M.S. Press.

Quinn, David Beers (2000) 'Thomas Harriot and the problem of America', in Robert Fox (ed.) *Thomas Harriot: an Elizabethan man of science*, Aldershot: Ashgate, 9–27.

Ramus, Petrus (1569) *P. Rami Scholarum Mathematicarum, Libri Unus et Triginta*, Basiliae: Eusebium Episcopium and Nicolai Fratris.

—— (1590) *The Elementes of Geometrie: written in Latin by that excellent scholler P. Ramus*, trans. Thomas Hood, London: Thomas Hood.

Randolph, Thomas (1630) *Aristippus, or the Iouiall Philosopher*, London: John Marriot.

Rathborne, Aaron (1616) *The Surveyor in Foure Books*, London: W. Burre.

Rattansi, P.M. (1972) 'The social interpretation of science in the seventeenth century', in Peter Mathias (ed.) *Science and Society 1600–1900*, Cambridge: Cambridge University Press, 1–32.

Ravenhill, William (1983) 'Christopher Saxton's Surveying: an enigma', in Sarah Tyacke (ed.) *English Map-Making 1500–1650: historical essays*, London: British Library, 112–18.

Record, Robert (1551) *The Pathway to Knowledg*, London: R. Wolfe.

Richeson, A.W. (1966) *English Land Measuring to 1800: instruments and practices*, London: Society for the History of Technology and Cambridge, Massachusetts: MIT Press.

Robinson, Henry (1641) *England's Safety in Trades Encrease*, London: Nicholas Bourne.

Ross, Alexander (1642) *Mel Heliconium; or, Poeticall Honey, Gathered out of the Weeds of Parnassus*, London: W. Leak.

Rountree, Helen C. and Turner, E. Randolph, III (eds) (2002) *Before and After Jamestown: Virginia's Powhatans and their predecessors*, Gainesville, Florida: University Press of Florida.

Sanford, Rhonda Lemke (2002) *Maps and Memory in Early Modern England: a sense of place*, New York and Basingstoke: Palgrave.

Saville, J. (1969) 'Primitive accumulation and early industrialization in Britain', in R. Miliband and J. Saville (eds) *Socialist Register 1969*, London: Merlin Press, 247–71.

Schmidt, Benjamin (1997) 'Mapping an Empire: cartographic and colonial rivalry in seventeenth-century Dutch and English North America,' *William and Mary Quarterly*, 54.3: 549–78.

Schwartz, Seymour I. and Ehrenberg, Ralph E. (1980) *The Mapping of America*, New York: Harry N. Abrams.

Seed, Patricia (1995) *Ceremonies of Possession: Europe's conquest of the New World 1492–1640*, Cambridge: Cambridge University Press.

—— (2000) 'Caliban and native title: "This Island's Mine"', in Peter Hulme and William Sherman (eds) *The Tempest and its Travels*, London: Reaktion, 202–11.

Shakespeare, William (1986) *Complete Works*, ed. Stanley Wells and Gary Taylor, Oxford: Clarendon.

—— (1987) *The Tempest*, ed. Stephen Orgel, Oxford: Clarendon.

Shapin, Steven (1988a) 'The house of experiment in seventeenth-century England', *Isis*, 79.3: 373–404.

—— (1988b) 'Understanding the Merton thesis', *Isis*, 79.4: 594–605.

—— (1992) 'Discipline and bounding: the history and sociology of science as seen through the externalism–internalism debate', *History of Science*, 30: 333–69.

—— and Schaffer, Simon (1985) *Leviathan and the Air-Pump: Hobbes, Boyle and the experimental life*, Princeton and Guildford: Princeton University Press.

Shapiro, Barbara (1969) *John Wilkins, 1614–1672: an intellectual biography*, Berkeley and Los Angeles: University of California Press.

Sherman, William H. (1995) *John Dee: the politics of reading and writing in the English Renaissance*, Amherst: University of Massachusetts Press.

Shirley, James (1962) 'The Lady of Pleasure', in A.S. Knowland (ed.) *Six Caroline Plays*, London: Oxford University Press.

Silverman, D.J. (2003) '"We Chuse To Be Bounded": Native American animal husbandry in colonial New England', *William and Mary Quarterly*, 60.3: 511–48.

Skelton, R.A. (1952) *Decorative Printed Maps of the Fifteenth to Eighteenth Centuries*, London: Staples Press.

Slotkin, Richard (1973) *Regeneration Through Violence*, Connecticut: Wesleyan University Press.

Smith, Captain John (1612) *A Map of Virginia: with a description of the countrey, the commodities, people, government and religion*, Oxford: Joseph Barnes.

—— (1624) *The Generall Historie of Virginia, New-England, and the Summer Isles ...*, London: Michael Sparkes.

—— (1986) *The Complete Works of Captain John Smith (1580–1631)*, ed. Philip L. Barbour, 3 vols, Chapel Hill: University of North Carolina Press.

Spufford, Margaret (ed.) (1995) *The World of Rural Dissenters, 1520–1725*, Cambridge: Cambridge University Press.

Spurr, John (1998) *English Puritanism, 1603–1689*, Basingstoke: Macmillan.

Stavely, Keith W.F. (1987) *Puritan Legacies*, Ithaca: Cornell University Press.

Stearns, Raymond Phineas (1946) 'Colonial fellows of the Royal Society of London, 1661–1788', *William and Mary Quarterly*, 3.2: 208–68.

Steggle, Matthew (2004) 'Aylett, Robert (*c.*1582–1655)', *Oxford Dictionary of National Biography*, Oxford University Press. Available oneline at: http://www.oxforddnb.com/view/article/932 (accessed 22 October 2004).

Stevenson, Matthew (1662) *Florus Britannicus; or, an Exact Epitome of the History of England*, London: T. Jenner.

Strachey, William (1953) *The Historie of Travell into Virginia Britania (1612)*, ed. Louis B. Wright and Virginia Freund, Hakluyt Society Second Series 103, London: Hakluyt Society.

Suckling, Sir John (1648) *A Collection of all the Incomparable Peeces Written by Sir John Suckling ...*, London: Humphrey Moseley.

Sullivan, Garrett (1998) *The Drama of Landscape: land, property, and social relations on the early modern stage*, Stanford: Stanford University Press.

Tawney, R.H. (1912) *The Agrarian Problem in the Sixteenth Century*, London: Longmans & Co.

Taylor, E.G.R. (1934) *Late Tudor and Early Stuart Geography 1583–1650*, London: Methuen.

—— (1954) *The Mathematical Practitioners of Tudor and Stuart England*, Cambridge: Cambridge University Press.

Taylor, John (1630) *All the workes of Iohn Taylor the water poet*, London: James Boler.

Taylor, Silvanus (1652) *Common-good; or, the Improvement of Commons, Forrests, and Chases, by Inclosure*, London: Francis Tyton.

Thirsk, Joan (1992) 'Making a fresh start: sixteenth-century agriculture and the classical imagination', in Michael Leslie and Timothy Raylor (eds) *Culture and Cultivation in Early Modern England: writing and the land*, Leicester: Leicester University Press, 15–34.

Turner, G.L.E. (1983) 'Mathematical instrument-making in London in the sixteenth century', in Sarah Tyacke (ed.) *English Map-Making 1500–1650: historical essays*, London: British Library, 93–106.

Tyacke, Sarah (1983) 'Introduction', in Sarah Tyacke (ed.) *English Map-Making 1500–1650: historical essays*, London: British Library, 13–18.

Underhill, Captain John (1638) *Newes from America; or, a New and Experimentall Discoverie of New England ...*, London: Peter Cole.

—— (1897) *Newes from America; or, a New and Experimentall Discoverie of New England ...*, in Charles Orr (ed.) *History of the Pequot War: the contemporary accounts of Mason, Underhill, Vincent and Gardener*, Cleveland: Helman-Taylor Co.

Vaughan, Henry (1678) *Thalia Rediviva: the pass-times and diversions of a countrey-muse in choice poems on several occasions*, London: Robert Pawlet.

Veen, Otto Van (1996) *Amorum Emblemata, Figuris Aeneis Incisa Studio Othonis Vaeni Batauo-Lugdunensis*, Antuerpiae: Henrici Swingenij.

Verner, Coolie (1968) *Smith's Virginia and its Derivatives: a carto-bibliographical study of the diffusion of geographical knowledge*, Map Collectors Series 45, London: Map Collectors Circle.

Vincent, Philip (1897) *A True Relation of the Late Battell fought in New England ...*, in Charles Orr (ed.) *History of the Pequot War: the contemporary accounts of Mason, Underhill, Vincent and Gardener*, Cleveland: Helman-Taylor Co.

Virgil (1961) *The Works of Virgil Translated by John Dryden*, trans. John Dryden, London: Oxford University Press.

Vitruvius, Marcus Pollio (1960) *The Ten Books on Architecture*, trans. Morris Hicky Morgan, New York: Dover.

Waller, Edmund (1893) *The Poems*, ed. G. Thorn Drury, 2 vols, London: Spence and Bullen, and New York: Charles Scribner's Sons.

Wallis, Helen (ed.) (1985) *Raleigh and Roanoke: the first English colony in America, 1584–1590*, London: British Library Publishing.

Waselkov, Gregory A. (1998) 'Indian maps of the colonial southeast: archaeological implications and prospects', in G. Malcolm Lewis (ed.) *Cartographic Encounters: perspectives on Native American mapmaking and map use*, Chicago: University of Chicago Press, 205–21.

Watt, Ian (1957) *The Rise of the Novel: studies in Defoe, Richardson and Fielding*, London: Chatto and Windus.

Watt, Tessa (1991) *Cheap Print and Popular Piety 1550–1640*, Cambridge: Cambridge University Press.

Webb, Stephen Saunders (1979) *The Governors-General: the English army and the definition of empire, 1569–1681*, Chapel Hill: University of North Carolina Press.

Weber, Max (1930) *The Protestant Ethic and the Spirit of Capitalism*, trans. Talcott Parsons, London: Allen & Unwin.

Webster, Charles (1975) *The Great Instauration: science, medicine and reform, 1626–1660*, London: Duckworth.

—— (1979) *Utopian Planning and the Puritan Revolution: Gabriel Plattes, Samuel Hartlib and Macaria*, Research Publications of the Wellcome Unit for the History of Medicine 2, Oxford.

White, John (1957) *The Birth and Rebirth of Pictorial Space*, London: Faber and Faber.

Whitney, Geffrey (1586) *A Choice of Emblemes and other devises for the moste parte gathered out of sundrie writers*, Leyden: C. Plantyn.

Wilkins, John, Bishop of Chester (1648) *Mathematicall Magick; or, the Wonders that may be Performed by Mechanicall Geometry*, London: Sam Gellibrand.

Winthrop, John (1846) 'General considerations for planting in New England', in Alexander Young (ed.) *Chronicles of the First Planting of the Colony of Massachusetts Bay, from 1623 to 1636*, Boston: Little and Brown, 270–8.

Wither, George (1635) *A Collection of Emblemes, Ancient and Modern*, London: R. Milbourne.

—— (1872–77) *Miscellaneous Works*, 6 vols, London: Spenser Society.

—— (1880) *Britain's Remembrancer*, London: Spenser Society.

Wood, William (1634) *New England's Prospect*, London: Iohn Bellamie.

Woodcock, Matthew (2001) '"The World is Made For Use": theme and form in Fulke Greville's verse treatises', *Sidney Journal*, 19.1/2: 143–59.

Worlidge, John (1675, 2nd edn; first published 1669) *Systema Agriculturæ: the mystery of husbandry discovered*, London: T. Dring.

Worsop, Edward (1582) *A Discoverie of Sundry Errours and Faults Daily Committed by Landemeaters*, London: G. Seton.

Yates, Francis A. (1964) *Giordano Bruno and the Hermetic Tradition*, London: Routledge and Kegan Paul, and Chicago: University of Chicago Press.

—— (1966) *The Art of Memory*, London: Routledge and Kegan Paul.

—— (1969) *Theatre of the World*, London: Routledge and Kegan Paul.

Index